LOW BRIDGES
and
HIGH WATER
on the
New York State Barge Canal

Charles T. O'Malley

LOW BRIDGES & HIGH WATER
ON THE
NEW YORK STATE BARGE CANAL

ISBN 0-9628533-0-5
1st & 2nd printing
Diamond Mohawk Publishing

ISBN 978-0-925168-38-2
3rd Printing
North Country Books

Printed in the United States of America

North Country Books, Inc.
311 Turner Street
Utica, New York 13501

This Book is Dedicated to
Muzz Gorney
Capt. Jack Maloney
and
Capt. Bob Gordon

LOW BRIDGES AND HIGH WATER

CONTENTS

ACKNOWLEDGMENTS

To give this book a good balance between yesterday and today I needed to make a few trips on a working tug. It would have to have a Captain with a tolerance for someone being underfoot and asking lots of questions.

The owner who came thru for me was Marty Kehoe, owner of the "Erin Kehoe." He couldn't have made a better choice than Captain Bob Gordon and his crew. I am forever in their debt. I received a bonus from this association. Bob and Eileen Gordon became good friends.

I have a special place in my heart for T. Emmett Collins, Dick Garrity, Rose Feeney, Tom Feeney and Grace Hancox (Matton Chapter) mother of my best friend from childhood, Bob Hancox. Their contributions were vital.

Graphics are very important to my book and I wish to thank Rod Stafford for his work and the family of the late Don Curtice for his. Also, many of the fine photos in the book are from the collections of Gene Baxter and George Michon.

A manuscript will never see the light of day without the skills of a good text editor. I was fortunate in having one of the best, Lionel D. Wyld. Lionel fulfilled another role. Many times he kept me going when I was ready to chuck the whole project. I am very grateful for his support.

PREFACE

In one respect, the transportation industry is no different from all others. To survive, it must keep pace with need, competition, and technology. Need is obviously the most important factor. The most modern, up-to-date industry is doomed to failure if its services or products are outmoded. An obvious statement? Yes, but how many times do we ignore or refuse to see this and continue, nonetheless, to pour money into a project long after the need exists? A case in point may well be the New York State Barge Canal.

As I write this, I see the canal system advancing toward oblivion at an alarming pace. As a commercial waterway, it will probably pass into history in my lifetime. This thought deeply upsets me, for my nostalgic romanticism forces me to protest. As futile as a child's excuses to stay up past bed time, I grasp for reasons why the canal should be kept alive. I may fail, but I offer this book at least as my eulogy to its passing.

The story begins in 1791 with the Inland Lock Navigation Company's canals, whereby natural streams in New York State were improved by private enterprise to begin a transportation route across the state. Next comes the era of the original Erie Canal, to the present as we see the Barge Canal traffic continue to decline. We are always reluctant to accept the end of an era. On the oceans, sail gave way to steam, which in turn was replaced by diesel. And now the canal boat has been very nearly wiped out by trains, trucks, and pipe lines. A token amount of each type is usually left to fill the minds of some with false hopes that their choice of the perfect transportation system will reign supreme again. Sadly, it never seems to happen.

I think change is more traumatic in the transportation industry because of its image of glamour. Ask a few lads what they would like to be when they grow up. I doubt if any would say "punch press operator," but becoming an airplane pilot or truck driver might appeal to them.

Twice a day while driving to work, I crossed the Erie division of the Barge Canal and followed along its banks for a couple of miles. Driving the same route every day, I thought how little we are aware of the objects we pass. We seem programmed and partially hypnotized, not even remembering anything between home and job. The absence of something, rather than the object, will often snap our minds into sharp focus. It happened to me late one summer. It struck me that I hadn't seen any tugs and barges on the canal that spring or summer. I found the answer when I started my research. Traffic had definitely fallen off.

Other years I always stopped my car and waved to the crews as I had done since I was a child living next to the Hudson River. Talking about it afterwards brought back all the memories of my childhood, when I played on abandoned tugs near my home in Troy, New York. I even remember the times my brother, Bob, ran away from home by hitching a ride on a tug. He usually lost his nerve by the time the tow reached Lock 7 on the Western Canal.

I was saddened by the the thought that the canal could be dying and no one was recording its demise. Someone jokingly suggested I do something about it. I'm not a historian, yet the suggestion that I write about the canal kept nagging me, so I decided to give it a try.

Much thought went into selecting the title. It is not meant to be cute. Low bridges and high water are an occupational hazard for the towing companies operating on the New York State Canal System.

Like all forms of transportation, the equipment (tugs and barges, in this case), are designed and built to the physical limitations of the environment in which they are to operate.

Length and width of the locks on the canal are fixed. Depth of the water and clearance under the bridges can vary.

If the channel fills with mud, it can be dredged. If low water occurs, bulk products can be reduced in the barges and water ballast pumped out of the tug to clear the bottom. In the event of high water, nothing can be done but to wait it out. This could take a few hours or run into days and even weeks. Delays are costly. Towing companies sell one product – time.

After selecting a title, the most important question must be answered. What is the value of this book? in recent years, I've come to realize most people consider the old Erie Canal has historical significance, but not the Barge Canal. This is reflected in a paucity of data in the archives of the Canal Society of New York State, the Canal Museum at Syracuse, New York State Historical Society, and most other historical societies of the state. The only book I found on the subject was *Canal Boatman* by Richard Garrity of Tonawanda, New York.

The historical value lies in the fact that we have, in essence, a cottage industry that has survived far into the Twentieth Century. Even the giants of today (Moran and McAllister) are still owned and operated by the original families.

Most of the companies have disappeared without a trace. There are no records or people alive to tell their story. The remaining companies, with one exception, are based in the New York City area. Of the ones I have written about, I chose one from New York City, four from Upstate, and one that first operated in New York City and is now based Upstate.

My choices are also colored by the need to cover the time period 1915 through 1984 and to show one-man organizations and at least one large one. I also wanted to write about some companies that not only ran tugs and barges, but also built them.

From the start of World War I until 1922, the United States Government controlled New York's canal system. Next came the 1920's when the canal was dominated by individual boat owners. Slowly these boat owners were replaced by companies operating three or more tugs and barges. The pattern changed as canal traffic declined and more and more boats were operated by fewer and fewer companies.

The canal went from a peak in 1922 of 1300 registered vessels to 143 registered in 1983. It is important to remember that even though there were 143 vessels registered, very few actually made trips on the canal. The same holds true to some extent with the 1300 figure, but the percentage running was higher. The traffic has declined to where the western end (Syracuse to Buffalo) has become a ditch for farmers (for irrigation), local fire departments and the movement of pleasure boats. There is a small amount of petroleum products moving on the Champlain Branch, a barge or two a week hauling tar to Lyons, New York and a few movements on the Oswego Branch.

The chapters about my trips on the ERIN KEHOE should give my readers a feel for the day-to-day operations of tugs and barges that remain today.

The following questions will be answered:

How could a meal of pancakes and sausages trigger a mutiny?

Why did Captain Shin Roberts run his tugboat on his knees?

What international company operated its own tugs and barges to move product and raw materials?

Why would Tom Coyne stop a loaded tow to pick flowers?

Is there a real difference between barges built by Feeney and Matton?

Why was the canal doomed before it was even completed?

What is a double-ender?

The Erie Canal became the New York State Barge Canal in 1915; from then on the Erie became a branch of the Barge Canal. It is also called the Western Canal and runs from Waterford to Tonawanda. The Champlain Canal is sometimes called the Northern Canal. The Oswego Branch does not have a second name, nor does the Cayuga and Seneca Canal.

CHAPTER 1

BEGINNINGS

It was early on a July morning in 1941. The exact day doesn't matter, even if I could remember. The day will very quickly become hot and sticky. The kind of day that leaves you breathless and in a slight stupor. Summers are like that in Troy, New York. I don't feel it, sense it or care at all. I'm 17 and headed for the greatest job in the world. I am going to be a fireman on a tug boat. I had been hired the day before for an opening on the old steam tug MATTON 10, owned by John E. Matton & Son of Cohoes, New York, whose boatyard was directly across the Hudson River from my home. She was built at Athens, New York, in 1903. The MATTON 10, 70 feet long, with a 100-horsepower engine, was old and weary when I met her.

I had dreamed for a long time about working on the boats. Dreaming gave way to reality, which in a short time turned to disillusion as I arrived on the dock to start my career. That career lasted one trip from Cohoes, down the Hudson River to pick up a barge of gasoline in mid-channel and haul it to Tonawanda. Fear was my overwhelming emotion that first day. It started with my family telling me all the horror stories about the rough, hard drinking, immoral boatmen. Fear that I couldn't learn the job, or would not be strong enough to do it. The size and ruggedness of some of the crew on my first boat rather intimidated me. I was six-feet-two and built like a needle.

As a junior fireman I would be on the 12:00 a.m. to 6:00 a.m. watch, twice a day. Yes, we worked twelve hours a day, seven days a week. The Captain, Chief Engineer, senior fireman and senior deckhand worked the more desirable hours of 6:00 a.m. to 12:00 p.m. We were also separated in another way. The deck force, Captain, Mate, the two deckhands slept in a room aft of the wheelhouse. The black gang (engineers and firemen), plus the cook, lived in the very aft end of the deckhouse.

Bunk assignments were made according to job status and seniority. The room was very small, with headroom less than six-feet-six, with two bunks on the port side and three on the starboard. The best one was the lower one on the port side, and it went to the Chief Engineer. The next best was the top one on that side, and it went to the cook. The entire starboard side was a disaster, due to cramming three bunks where only two should have been. The middle one was the best of the lot, and it went to the Assistant Engineer. Next came the top one for the senior fireman. Last

and worst went to me, Charlie. To add insult to injury, the Assistant weighed in at about 270 pounds. With his displacement, my space was reduced to where, even with my needlelike body, I had more than a slight problem getting into bed. With my bunk just six inches off the deck, I had to lie flat out on the deck and slide laterally into the bunk. Once in, I could not turn completely over. My first thought was, what would happen if the hippo above me broke the bunk? I'd be mashed potatoes!

The MATTON 10 did have one improvement over some of the tugs of the same age. Instead of having to use a bucket in the engine room, we had an enclosed toilet of miniscule size. So small, in fact, you couldn't walk into it, you had to back in. I often wondered how our hippo ever managed this feat. Our bathing facilities were equally modern, consisting of a shower head (without enclosure) in the engine room. It was almost impossible to stay clean on a coal-burning tug, but it was extra important, because this was in the days before deodorants. To wash our clothes, we put them in a bucket and inserted a small steam line in to boil the water.

Dividing my sleeping time in two parts took a little getting used to. After one day, total exhaustion solved that problem very nicely. On the other hand, eating four meals a day, while new to me, certainly didn't offer much of a challenge. Tugboating like this wasn't all bad. There were two distinct advantages of steam tugs over Diesels: they were much quieter and free from the intense vibrations found even today on Diesel tugs.

About 11:00 a.m. we threw off our lines, left MATTON's yard, and headed down river. I had my first meal aboard and went to relieve the first watch. I was so nervous I don't recall if I enjoyed the food or not. As I climbed down the ladder to the boilers, the first thing to strike me was the heat. I was positive I would never survive six hours in that hell hole. The Assistant Engineer (my boss) handed me a number 9 scoop (coal shovel), flung open a firebox door, pointed to the coal and said, "Get started!" He showed me the importance of throwing the coal, so that it spread evenly across the fire bed but not so thick that the air couldn't get through. This was vital, because the tug didn't have forced draft induction fans, although we did have a way to increase the draft by feeding exhaust steam into the stack. In any event, the thickness of the coal was critical.

He then introduced me to my next tool – a small machinist hammer. To maintain the water level in the boilers, you had to turn on a steam-driven feed pump and stick your head around a bulkhead to watch the water rise in the sight glass at the top of the boiler. Not difficult, except

MATTON 10. Note coal on deck and water barrel on bow. Photo by George Michon.

that it was necessary to gently tap the pump occasionally because it had a tendency to stick. Equally important was when to add water. If you waited too long, you add too much cold water and that knocks the steam pressure down. This, of course, slows down the engine, and that has a tendency to upset the Mate, who has the wheel on this watch.

I soon learned the key to being a successful fireman, not that I ever reached this point. It was keeping the pressure to a point just below where it will pop the safety valve. Blowing a safety valve was cardinal sin. It marks you as a novice, wastes steam and, more importantly, it brings on a startling aide effect. Bear in mind that a steam tug carries a mantle of coal dust and ashes from the stack aft to the fantail. The thickness is in direct proportion to the last time some idiot like me popped the safety valve. I do believe I set some kind of record, in the time it took me to let this happen.

A few minutes into the watch and we were entering the first and only lock between Troy and New York. I forgot lesson number three and didn't slap on the hunk of steel that acted as a damper. The safety valve let go. When that happened, I dislodged somewhere in the neighborhood of 37 tons of accumulated coal dust and ash. To make matters worse, some of the crew, including the Captain, were standing on the fantail, digesting their lunch and trading gawks with some onlookers near the lock wall. The language would have done a New York cabbie proud. Like Superman, I made the deck without touching a rung of the ladder. You would think I would have been smart enough not to rush to the scene of my crime, but no, I had to survey my handiwork. The Captain nailed me to the bulkhead with his eyes. I froze and fully expected to be thrown overboard. All the horror stories I'd ever heard about the roughness of tug crews strangled my thoughts. With tremendous control and through gritted teeth, Captain Napoleon ("Poley") Minor informed me that they don't sail tugboats in the sky and in the future would I try to keep the steam in its proper place. With my face beet red, I turned and dove for the sanctuary of the boiler room.

"Balance" best describes the key element in operating steam boilers. The amount of steam pressure needed is directly proportional to the demand of the engine. Accomplishing that feat is what makes life interesting for a novice fireman. If you lessen the demand for pressure, the pressure builds up very quickly and you can't turn steam off as you would a light switch. I knew I had to place the damper plate over the opening below the grates in order to cut the flow of air to the fire. The part I hadn't learned was when to anticipate the correct moment. Once inside the lock, the pressure has no place to go but up and out. That's the way the system worked on old tugs like the MATTON 10.

For the rest of my watch I never left the boilers. I was terrified I'd make another mistake. Meanwhile, something else was taking place in the boilers. As you know, as coal burns it creates ashes that have to be disposed of. Most people in the 1940's had a coal furnace in their homes, and I naively thought the ashes were removed from a tug boiler in the same way they were in a home furnace. But was I ever wrong! A home furnace has a movable grate, and you simple turn a handle or move a lever back and forth to let the ashes drop through the grates to the ash pit below. On a tug the grates are individual pieces that are fitted so that you can replace one grate at a time, without shutting down the boiler. Front to rear, there are two rows of them. Removing the ashes was a hell of a lot more difficult on the MATTON 10. The first step was to take a long heavy tool called a "spud" and fold one half of the fire on top of the other half. This leaves the ashes exposed, so you can then use a long handled hoe to rake them out on the cement deck at your feet.

The air temperature is high in a boiler room under any circumstances; now it becomes unreal as the ashes pile up. You don't realize the heat is being absorbed by your pants as they hang loosely away from your legs. You are forcefully reminded of it as you step back and your pants touch your bare legs. The only way I can describe the feeling is to ask you to imagine someone using a steam iron to press your pants while you are still in them!

The *coup de grace* to my health and well being came next. Those ashes have to be cooled down before shoveling them into a steel bucket, to be raised by the deckhand and dumped over the side. The water hit them and immediately engulfed me in a dense sulfurous cloud that tears at the nose and throat and threatened to asphyxiate me on the spot. I gagged, and stepped back until the worst was over. Three more times I endured this purgatory, because I had to fold the top back onto the side I had just cleaned and go through the process again – plus the other boiler. There is a final touch of skill required with this chore. With the fire doors open and half the grate exposed, the pressure will drop quickly and the engine will slow down, so speed is of the essence. Oh, for the glamorous life of a boatman!

As bad as all this sounds, the worst was yet to come. I mentioned before that grates have to be replaced. This is the ultimate test of fortitude. How you change them is the type of nightmare that old timers relish telling the new kid. You don't really believe them and you're not 100 percent convinced. On the third day I was convinced. I opened the door on the starboard boiler and saw a suspicious black hole in the fire bed. I called the Engineer and he confirmed my suspicions. Naturally, the spare grates were not in an easy place to get at. They were in the lazarette, the tiny storage place between decks, with hundreds of other spare parts.

The first step was to clear the fire around the broken grate. Then you use a special spud that for all the world looks like a giant toothpick, to jiggle the pieces so they fall through to the ash pit. This was the easy part. With the new grate in your hands, you extend your arm into the fire box, fully expecting it to incinerate, and pray you drop it in the right spot the first time. Mind, while you're in this position, your face is about three inches from the inferno.

I'll let your imagination take over to visualize what it was like to change a rear grate.

The rest of the trip becomes a blur in my memory, but I recall the food and the heat of the galley: the huge, black monster of a range, that gave off hot breath and was never shut down. The same is true today, except that the fuel is oil instead of coal. We ate in a bath of sweat, but I never complained about it, for very good reason. We were just coming out of the Great Depression and the memory of it was fresh in my mind. My Mother died when I was twelve, and after that my Dad, brother and I ate our own cooking. I remember that fresh bread was a real treat. Anyway, the food on the MATTON tug was good, plentiful and filling.

I had one final run-in with the Captain. I can't blame what happened on immaturity, for it was plain stupidity. I went to the bow of the barge to have a cigarette after my watch. I sat near a vent for the gas fumes and lit up. The Captain didn't need any mechanical amplification of his voice to reach my ear. I deserved every syllable of every word he used.

By the time we reached Tonawanda I was really down. I felt I was the biggest *klutz* in the world. I would never master the art of firing a boiler. After supper, while the barge was being pumped out, the entire crew except me headed for the nearest tavern for some relaxation. Again, as I look back, I think if just one person had stayed and talked to me, I might still be on the boats. I was homesick and feeling sorry for myself, so I took off and started to hitchhike back to Troy.

There is one postscript to this beginning. In a manner of speaking, the paths of the Captain and I crossed ten years after I left the boat. I had moved to Rochester and one night in the paper I spotted a small article about a tugboat Captain who had been shot by a kid while the tug was passing through the Rock Cut in Rochester. Thank God, it wasn't serious. But I didn't have the courage to face him; I figured he's still hate me for running away.

Lock 2, Waterford, New York. Motorship OSWEGO SOCONY, 1940.

CHAPTER 2

THE BIRTH AND EARLY YEARS OF THE BARGE CANAL
Information for this chapter is based largely upon
History of the New State Canal by Noble E. Whitford (1921).

Conception and Ground Work

It's surprising that the New York State Barge Canal, as we know it today, was ever built. The magnitude of the obstacles that had to be overcome seem, even in retrospect, to be enormous. The general public's attitude, ranging all the way from indifference to the more vocal call to do away with the waterway entirely, didn't help. Probably the granddaddy of all obstacles to the Erie and all other canals were the railroads, arch enemies of all inland water transportation systems; and yet another obstacle was a $9,000,000 improvement fiasco in 1894-1895. It must be remembered that changes in the canal system were always preceded by years of proposals, counter proposals, studies, surveys, and political procrastination. In June 1884, the annual meeting of the American Society of Civil Engineers, with participation by a score of prominent engineers, took up discussion of the canal questions. Elnathan Sweet, State Engineer at the time, led the discussion in presenting a paper on the subject of building a ship canal across the State. This idea was one that always seemed to be around. It had been proposed many times, and would be again, even as recent as 1979. In 1884, no action was taken by the State as a result of the ASCE meeting, but the engineers probably did have the beneficial effect of helping to keep public opinion in favor of canal improvements.

Six years later, in 1892, an act of the Legislature authorized the election of delegates to a Constitutional Convention to meet in 1894. One of the delegates' duties was the consideration of amendments relating to the care and improvement of New York State canals. To enable the canals to compete with rival transportation systems, it became obvious that canal channels would have to be made larger. The first proposal took the canal depth from seven feet to nine, which canal advocates figured would meet the need of water transportation for a number of years. Two or three years earlier, however, a plan very much like this one had been proposed by State Engineer Martin Schenck as the practical canal of the future. His official report (1892) contains the first such presentation of a plan of enlargement which in general, closely resembled the New York State Barge Canal of the present day. In this same report, Schenck called attention to a set of conservative resolutions adopted by a canal convention held in Buffalo on October 19, 1892, relative to enlarging the canal prism. Since the demand for a larger canal seemed imminent, he recommended to the Legislature that funds be provided to enable him to make a survey of the Erie Canal from which he could make plans for the proposed increase from seven to nine feet. If the funds had been provided at that time, there would have been something better than the antiquated survey of 1876 on which to base an estimate of cost when the Constitutional Convention of 1894 called on the State Engineer and gave him only twelve days to prepare the estimate. Much of the trouble connected with the nine-foot enlargement might have been avoided had Schenck's report been acted upon prior to the Convention.

At any rate, the Convention called upon the State Engineer for estimates of cost for deepening the canal to nine feet. For several years, the State Engineers had been urging, with no success, that the legislatures appropriate money for a survey upon which to base such an estimate. The State Engineer in 1894, Cambell W. Adams, made what use he could of the old and entirely inadequate survey of 1876 to prepare an estimate within the allotted time. His estimated amount – only a guess assuming all conditions to be favorable – was $11,573,000, with an additional million needed for repairing and rebuilding walls.

New York State voters approved an amendment to the State Constitution in 1894, and the Legislature of 1895 passed a referendum for deepening the canal to nine feet, naming $9,000,000 as the amount to be expended. This sum was arbitrarily fixed by the Legislature without further consultation with the State Engineer. The measure was approved by popular vote at the general election. The statute required that work be begun within three months, the plans and estimates were prepared. The final probable cost was $13.5 million, or $15 million with engineering, inspection, and advertising added. A later revision would make the total $16 million.

Although only $9 million was available to spend, construction was begun. By cutting out certain pieces of work, the State hoped the cost would be within the amount of money available. By the latter part of 1897, however, it began to be realized, first by those connected with the work and then by the general public, that the proposed improvement could not be completed without a more adequate appropriation; and with these reports came rumors of alleged frauds and extravagance in the administration.

Although construction on the Barge Canal began in 1905, we have to go back to 1899 to see the initial efforts that made this idea a reality. On March 8, 1899, Governor Theodore Roosevelt appointed seven men to serve on what was known as the Committee on Canals. In contrast to the haphazard way the earlier and costlier improvement was done, this committee was very thorough in their investigation. The committee's job was to formulate a State canal policy. During the time of its investigation, there was so much uncertainty as to what the findings would be that there was little that pro-canal men could do except wait for their report.

One step, however, the canal advocates could take and did. A call was sent out in May 1899 to hold a State Commerce Convention. This convention served the purpose for which it was called into existence, to revive a discussion of the canal improvement question at a time when it appeared to be a lost cause. The convention not only revived the discussion, but it succeeded in arousing the support of hundreds of the most influential business and politicians in the State. In addition, members of the New York Produce Exchange proved to be strong canal advocates. During the summer and fall of 1899, they held a series of meetings that resulted in the adoption of a resolution favoring the construction of a canal of a depth of not less than 14 feet and corresponding width, with a new alignment of canal as necessary, by canalizing rivers – an important proposal.

One member of the Committee on Canals visited Europe and made a study of the canals of France, Belgium, and Germany. For the welfare of the canals, a committee such as this proved indispensable in overcoming the people's apathy, and bringing about a willingness to make whatever improvement seemed best. With the public bewildered and distrustful, it was necessary that their leaders and advisors should be those in whom they could have implicit confidence. The personnel of Roosevelt's committee, together with the Governor's well-known reputation for straight-forward dealing, furnished ground for confidence. The report, when it was presented, proved to be unbiased and authoritative.

On January 15, 1900, the Committee on Canals sent its report to the Governor. It recommended that the Erie, Oswego, and Champlain canals should not be abandoned, but should be maintained and enlarged, and the Black River and the Cayuga and Seneca canals be maintained as navigable feeders but not be enlarged at the time.

Ten days after the Committee on Canals presented its report, the New York Commerce Commission submitted its report to Roosevelt. (This commission had been appointed by Governor Black in 1898 to inquire into the causes of decline in New York State Commerce.) In essence, the

Commerce Commission report recommended that the canal system not be abandoned, that a ship canal not be built, and that the canals should be enlarged. In addition, a portion of the report helps us to understand how, despite previous setbacks, the people were willing to authorize another, much larger expenditure. The report stated that there never had been sufficient authority, or indeed any authority, for thinking that $9,000,000 (as earlier voted upon), would be enough to complete the work.

Before the legislation of 1898 ordered the appointment of a commission to investigate the $9 million dollar improvement, it had before it two other measures. One was a referendum bill for raising $7 million to complete the work in progress; the other was a proposition – a recurring theme – to turn over the canals to the Federal Government. This latter measure created a storm of protest, and, when it came to a vote, was defeated by a large margin. The bill to raise $7 million never passed. Canal advocates figured it wise to let the matter drop, and the $9 million dollar improvement was never completed.

Apart from the physical changes recommended by the Commerce Commission report of 1900, there was a subject that in retrospect may have contributed a great deal to the less-than-hoped-for success of New York's canal system: A change was wanted in the corporation laws.

As a result, an amendment was added to the law governing transportation corporations which stated that the stock in companies operating on the canal could not exceed $50,000, and the minimum was lowered from $20,000 to $5,000. The reason behind this amendment seems to have been to give a competitive break to the small boatman whose continued existence appeared precarious. Large navigation companies were quickly gaining control of elevators, terminals, and waterfront property; and the consequently excessive terminal charges were driving traffic from the canal. Although the intentions may have been noble, in reality the amendment prevented the formation of companies large enough to either command the cooperation of connecting transportation lines or to carry on their business in an efficient manner. There was also a feeling that having to deal with individual boatmen accounted for much of the non-use of the all-water-route by Great Lakes shippers. The $50,000 limit was eliminated, but a sentence was added to try to retain some protection against abuses. It said, "No railroad corporation shall own or hold stock in any such corporation."

During 1903, there were many counter bills and schemes to scuttle the canal enlargement. Enemies of the canal, for example, adopted a regulation proposing to eliminate the constitutional prohibition against selling the canals. It provided, further, that the canal bed not be used for a railroad, but to be used exclusively to haul freight. Fortunately, this effort failed even to get out of committee.

A former State Senator proposed to form a corporation that would carry freight from Buffalo to New York City in a third to a half of the time required by canal boats and at a cost no more than that on the existing canal. His plan to build an electric or steam railway was too much of a pipe dream to worry canal men. Just how the new railroad was to do so much better than the existing New York Central in both rates and travel time was not explained.

Another proposition, presented to a joint meeting of the Senate and Assembly canal committees by the International Towing and Power Company, involved electric towage and a plan to build on the outer side of the tow path a steel structure that would carry rails on which two electric tractors could run. These would be able to pass each other in opposite directions and be so located as to not obstruct the towpath which might still be used by horses or mules. International Towing and Power Company estimated that boats could be towed from Buffalo to Albany for 50¢ per ton and that electric equipment would cost about $7,500. Incidentally, at that time, boats were being towed by steam canal boats for 50¢ a ton from Buffalo to New York, 150 miles farther than the proposal offered.

Anyway, after much debate in the Senate and Assembly, the bill for a Statewide canal system finally passed. On April 7, Governor Benjamin B. Odell, Jr. gave his approval to the canal referendum, and it became Chapter 147 of the laws of 1903. The estimated cost was $101,000,000, and the final victory was by no means won with the signature of the Governor. The campaign had only begun. The two opposing armies simply transferred their battlefields from the legislature to the electorate.

Other books and records are available on the details of efforts pro and con that went into the ensuing campaign to win voters to the respective sides. One event occurred just prior to the election that may have had an influence on the vote. The International Towing and Power Company had obtained permission during the summer to install a mile of track on the towpath at Schenectady, and on October 26, a public demonstration of electric propulsion was made with Governor Odell, other state officials, and boatmen. The hauling was done by a tractor, called an "electric mule." From a technical standpoint, the test was a success. It began with one, and worked up to four loaded boats of the maximum size in use at that time, 240 tons. They were towed at five miles per hour with no sign of bank wash. Then two electric mules hauled boats in opposite directions, showing how the boats could meet and pass one another. The company declared they were ready to equip the canal at their expense and operate under whatever terms the State dictated. The Superintendent of Public Works was inclined to regard the proposition favorably, but consent was withheld for the time being.

The anti-canal press was quick to exploit this event, stating that the demonstration proved the utter uselessness of spending a vast sum of money as called for in the referendum to create the Barge System. Several reasons are apparent for suspecting that anti-canal influence was behind this demonstration, for experiments in electric propulsion on the canals had been made in 1895 and 1903, always just *before* the people were to vote on a referendum for canal improvement. The timing of the experiments, within two or three days of an election, is too suspicious to be coincidence. Steel construction experts in the State Engineer's department did some estimating of their own and said the cost stated by the company was ridiculously low.

In any case, it was all academic: the referendum passed and the new canal system became a reality.

One large hurdle remained before construction could begin, involving the question of whether concrete or masonry should be used in construction. Engineers had little doubt which method to use, but there was tremendous pressure from stone cutters, bricklayers, and masons. Using concrete would be a death blow to them. Their case rested on the fact that the State hadn't used concrete to any great extent, and it was not in general use at that time.

It is pleasant to note that the canal authorities took considerable pains to show why concrete won out. Some failures of large concrete structures had occurred, to be sure, and there was some feeling that upstate New York's climate was too severe for the successful use of concrete. Thus, in 1904, an engineer was sent throughout the South and Midwest to talk with engineers who were experts in the use of concrete. He also examined structures as far north as Duluth, Minnesota, which is 275 miles farther north than the Barge Canal. He came back convinced concrete should be used. There was an added bonus in the final decision, for using stone masonry would have added over $16 million to the project! One significant change was, however, made after construction began. Enlargement plans called for the Cayuga and Seneca section and the Black River portion to be retained as feeders and not be enlarged. As had happened with the building of railroads (spur lines), everyone wanted a "piece of the action." The folks in the Finger Lakes area felt left out and argued for the enlargement of their canal. They cited the great potential for the shipment of cement from a new plant on Cayuga Lake and the bottomless pit of salt in the surrounding area. State legislators caved in and spent $7 million to fulfill their wishes, although, unfortunately, the promised volume of traffic never materialized. No campaign to enlarge the Black River Canal seems to have occurred. It remained simply a feeder, with all traffic ending on the Black River Canal in 1922.

Terminals

It seems incredible that it was eight years after the Barge Canal was authorized before state-controlled terminals were added to the project. New York had a long history of neglect in this respect, since many people, as well as the railroads, knew the importance of owning and controlling terminal facilities. Private owners controlled much of the waterfront in seaport, lake harbor, and river cities; in addition, a large part of the strategic sites were also controlled by the railroads. Naturally, it follows that these facilities were not operated for the benefit of water-borne traffic. Reading through the literature, one wonders why this fact seems to surprise so many people. Even the Federal Government, with a blue-ribbon commission, came to an obvious conclusion in 1910 when the National Waterway Commission conducted a survey. Only two ports in the United States – New Orleans and San Francisco – had established public control of terminals at all comparable to the municipal supervision existing in most European ports.

Compounding the problem of the railroads' control of much of the waterfront was the method of handling the cargoes once they reached a port. Noble E. Whitford's *History of the Barge Canal* (1921) comments succinctly on the subject. Whitford quotes from an article by Edward Mott Wooley in the January 1912 issue of *System:*

> In many places, trains of cars are carried on boats across rivers and straits, and long distances by water. There is no reason why grain, livestock, cotton, and much heavy freight which is costly to unload and which must take a part of its trip to market by rail, should not be carried on boats in the cars – the original shipping packages. Neither does there seem any reason why huge boxes each containing

a car load should not be made capable of being swung from the boat to the flat car to which it might be fitted and back to the boat again when necessary. The cranes capable of doing the work are already invented, and in use.

The deplorable conditions existing in marine traffic through the lack of machinery is seen when it is known that the terminal costs at New York and Liverpool exceed the hauling costs of the 3,000 mile voyage.

The human worker still reigns practically supreme on the docks in all his primitive wastefulness. He rolls up an annual payroll of millions, he congests traffic by his complex and cumbersome motions. He strikes when he pleases and ties up whole harbors.

Half the commerce of the nation comes through the Narrows and is distributed from the wharfs and piers in the vicinity of Greater New York. It comes in leviathans, but is seized upon by an army of human ants who spread themselves over the docks in a maze of inefficient and costly motion. . . .

In view of the immense volume of freight loaded and unloaded by vessels every day, the paucity of handling facilities, viewed from the standpoint of modern business management, is almost incomprehensible.

It is important to point out that Mr. Wooley's observations were made decades before we had piggyback railroad cars and container ships. In the early 1900's, meat shipped to London was off-loaded in Southhampton directly into wagons. The wagons were then placed on railroad cars and hauled to London, ready for the horses which hauled the wagons to the butcher shops.

The decline in the amount of freight carried on the canals was well known, but that this falling off was entirely in through-freight was a fact that was not fully realized until a study was made by the Terminal Commission. According to the Commission, the decline was due to a lack of independent terminal facilities at Buffalo and New York and to the increasing control by the railroads of terminals not only at Buffalo, but also throughout the Great Lakes regions. A contributing cause was the refusal by railroads to prorate on through routes where freight would naturally go part of the way by water. In 1917, the New York Legislature enacted a law aimed at the correction of this condition – an act to regulate joint rail and water routes.

To illustrate the railroad's action, the National Waterways Commission noted, "In many cases, the route which is the natural one, would be by water for three quarters or more of this distance, yet the charge for the remaining railroad haul is so considerable as to render carriage for the longer haul by water unprofitable."

Viewing the subject from a slightly different angle, the Terminal Commission said:

> The railroads have always refused to either prorate or through-rate with canal carriers, but on the contrary, have only been willing to receive freight brought to them by canal boat in the most unusual manner, such as forcing them to discharge their freight at places other than railroad wharfs, and then team it to the railroad wharfs, instead of allowing them to come directly to the wharfs. By refusing, on the other hand, to deliver freight to canal boats at their wharfs, they have been able to prevent them from carrying large quantities of freight that would otherwise have been shipped by the canals.

It is baffling why canal people should think the railroads would do anything to help a competitor!

In any event, $19,800,000 was appropriated in 1911 for terminal construction. The framers of the law foresaw the possibility of individuals or corporations trying to monopolize the terminals and did something about it. These terminals would remain the property of the State and be under its management and control forever.

At this point it appeared that the State of New York was really getting its act together and – even if a bit late – the State was trying to make the improved canal system a success. One important omission in the terminal law existed, for it was void of any mention of grain elevators. This seems a real shocker, because grain was from the very beginning of the Erie Canal the Number One commodity moved. Hence, the State may have been more than a little shortsighted in the planning department.

Proof of this observation lies in the Terminal Commission's excuse that they didn't have enough time to study the problem, yet beginning in 1919, the State Engineer made elevators an issue in his yearly report, but it wasn't until 1920 that the Legislature took any action. That year, the Governor, the State Engineer, and the Superintendent of Public Works made strong appeals to the Legislature for elevators at Buffalo, New York City, and Oswego. Finally, a bill was passed in 1920 to build elevators at Oswego and Gowanus Bay (New York, head of the connecting Gowanus Canal).

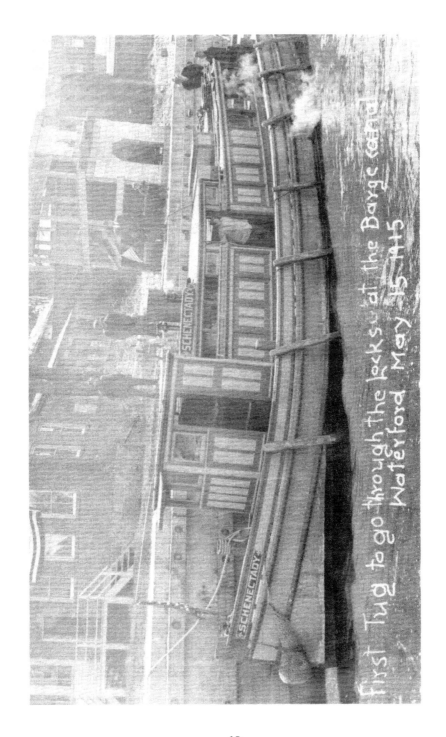

First Tug to go through the locks at the Barge canal — Waterford May 15, 1915

19

Buoyboats and state tugs. Waterford flight, circa 1918.

It's interesting to recall what the three State officials said to convince the Legislature to act. In 1880, New York canals carried more than 30 million bushels of wheat from Buffalo, using thousands of canal boats. In 1917, only one-half million bushels were carried in less than 400 boats. It was not because grain no longer came to Buffalo, far from it. Close to 100 million bushels were received in Buffalo every year, and at the close of navigation it happened that twenty times as much grain was found lying in Buffalo harbor awaiting shipment as the canal had carried its whole season.

The State had made excellent provision at its terminals to handle more other commodities, but no facilities for the grain traffic had been furnished. Grain normally amounted to 50 or 60 percent of all eastbound tonnage. Advocates of the Barge Canal affirmed that no other route, rail, or water could compete with New York waterways in this traffic if proper equipment was supplied.

The conditions under which it was necessary to handle any grain reaching New York by water were intolerable. For many years, there had been passing through that harbor half the foreign and domestic commerce of the entire United States and yet there existed but five or six grain elevators and only two were situated so that canal boats could reach them. Both of them were owned by railroads, and there was no chance for canal boats to be allowed to use them. As a result, barges loaded with grain had to wait the arrival of ships to which their cargoes could be transferred. This might take a week or two, sometimes longer. During the delay, demurrage charges mounted, the earning power of the barge was lost, and the cost of shipping grain by canal was rising prohibitively. The number of boats on the canal was notoriously inadequate for the tonnage the waterway could handle. But no matter how many barges might be in service or how much grain might be available for shipment, the continuation of such conditions in New York harbor could result only in wiping out the canal grain business and with it the very foundation of canal traffic. The State's final answer, for its own good, was to build an elevator at Oswego upstate and at Gowanus Bay in New York City.

Construction

The first actual contract work was first performed at Fort Miller on the Champlain Canal on April 24, 1905; the first work on the Erie Division was done at Waterford on June 7, 1905. In 1910, the first of the Barge

21

Canal locks was used: it was the lock at Baldwinsville. At the end of 1910, about one-third of the construction of the entire system had been completed.

Early in 1911 a break in the canal at Bushnell's Basin near Rochester, where the canal channel is carried on a high embankment, called attention to the need for some means of guarding against similar accidents. The contract plans were altered so as to provide for carrying the canal over the embankment in a concrete trough. This trough held up until 1974, when its collapse caused by a contractor working on a tunnel under the trough caused considerable damage to nearby houses.

In 1914, just before long stretches of river canalization were about to be opened to traffic, canal officials had to consider a new feature of construction. Aids to navigation had to be provided lights, buoys, and lighthouses. In the old canal, the channel ran generally in land line, bordered on both sides by immediately adjacent banks; in a few river sections, it hugged closely to the bank on which the towpath was built. In these older channels, a boatman had to be drunk or asleep to run aground, but in the new canal nearly three-quarters of which lay in canalized rivers or lakes, conditions were entirely different.

After much talk about color schemes for lights and buoys, the State finally adopted the Federal standards of red, white, and black. On returning from the ocean, red buoys and red lights are on one's right, and black buoys with white lights on the left. An easy way to remember is what the Navy teaches: RRR ("Red, Right, Returning"). By the time the Erie Division opened, more than 1500 lighted aids to navigation were in operation.

A far more important issue, lack of funds, was rearing its ugly head as early as 1912. State engineers were more than a little paranoid about this situation, because of the bad smell left over from the $9 million dollar fiasco cited earlier. Besides, their professional honor and reputations were at stake. They worked very hard to keep from over-running estimates. Even so, there were the usual increases in labor and materials, just as added costs have plagued every large construction job ever undertaken.

Most of the added costs in the Barge Canal situation were not under the control of the engineers. In 1905, for example, the Legislature amended the original Barge Canal Law, giving power to the Canal Board to increase the width of locks from 28 feet, as originally provided, to 45 feet. This added about $2,500,000 in costs without increasing the appropriation. In addition, both the Eight Hour Labor Law and the Workmen's Compensation Law were passed subsequent to the making of the estimate in 1903, and both meant additional cost of construction. Despite all obstacles, the cost of actual construction overran the estimate by only $6,000,000.

By the spring of 1915, an important section of the new channel, from Waterford to Rexford, was ready for opening. Because this was the portion that started at the Hudson River to begin the route across the state and also because it was an unusual section containing the greatest series of lift locks in the world, it was decided to make a gala occasion of the opening day. The ceremonies on May 15, 1915, consisted in the passing of a boat through the five locks at Waterford and on through he new body of water formed above the Crescent Dam. On board were Governor Charles S. Whitman, Secretary of State Francis M. Hugo, State Engineer Frank M. Williams, and many other dignitaries. A second boat carried the press corps.

Exactly three years later, on May 15, 1918, the new Erie Division of the Barge Canal was opened to navigation throughout its entire length.

In 1919, one hundred lights were added to the 1500 already in operation, and the State was successful in getting some boatmen to run at night. In this same year, a fleet of the old-style canal boats made a new record for passage from seaboard to Lake Erie in four and two-thirds days. In 1920, another 100 channel lights were added. Night travel on the canal increased greatly during 1921, and 400 more lights were added. Fleets then regularly made the trip from Buffalo to Troy in less than five days, and the round trip from Lake Erie to New York and return could be completed in 14 days.

State Towage

The Barge Canal marked the end of the century-old custom of animal towage on New York canals. No towing paths were provided on the new waterway, except such as were used temporarily during the period of transition. In 1914, two stretches of new channel which had no towing paths (Erie section) were opened to navigation. One lay in Wayne County, about twenty miles in length, and the other was a portion of the Mohawk River between Vischers Ferry and Rexford. Being thus isolated, with portions of the old canal extending from either end, it was necessary to provide some means of towing animal-drawn boats across these sections, since such craft constituted a large part of the shipping then in service. Accordingly, the Legislature appropriated $40,000 to pay for tugs to do the towing in these sections. In 1915, the Wayne County portion had been lengthened, but it was still isolated. The Mohawk River section had been extended

Unloading packet freight, Buffalo, 1919.

Tug "Alice R" and barges "Acacia" and "Jessie" at Schenectady, Lighterage Dept., General Electric Co.

to the Hudson River at Waterford and therefore had not the excuse of isolation in requiring State towage, but another appropriation was made that year and the practice was continued. On the Champlain and Oswego canals, boatmen had adapted themselves to the new conditions and had their own facilities for towage; but on the Erie branch, sixty percent of the boats plying on the canal were horse-drawn and again the State had to furnish tugs. Each year after 1914 and until 1921, except during 1918 when the Federal Government was in full control of canal transportation, the State made provision for towing such boats as had no other means of propulsion.

It doesn't take anyone very long to discover that it was a huge and costly headache for the State. First, the State had no expertise in running what was essentially a medium-size towing company spread out all over the State. To make matters worse, they had no real day-to-day control over the movement of the tugs. The removal of the towpath was a traumatic experience from which most of the boatmen would never recover. The majority couldn't make the transition, but at the time they thought they saw a way out of the dilemma, for they did try to get the State to continue towing. It was reasonable for the State to provide towing for a couple of years until the boatmen made the conversion to power. Until 1917, the mileages and the costs were small, but then the State began to charge 20 cents per mile per boat. This rate turned out to be 50% of what it should have been for the State to break even.

Apart from the money, delays made towing unworkable. The canal was divided up into 16 sections with one tug assigned to each section. Since a tug needed 24 hours to make a round trip, delays became a way of life. If a tug was missed by a few minutes, the wait was 24 hours. Inevitable breakdowns just increased wait times. Worse still, tugs refused to run at night, so the towing system operated on a 12-hour day. Interesting enough, though, the tugs would run at night if it suited their purpose, such as a good place to tie up or in an emergency. Blame could be laid on the Superintendent of Public Works who intended that the tugs should run around the clock without his seeing to it that the specifics were spelled out in the contracts. Hiring competent and trustworthy crews also was proving difficult.

By the 1920 season, the Superintendent was having a rough time. The money allotted was totally inadequate as labor, and the price of coal had increased greatly. In addition, traffic was up more than had been anticipated.

Shippers were aggravated by the delays, but there was another more pressing problem. It was reasonable to assume that with an enlarged canal, shippers would be looking to build larger barges, and they did. Unfortunately, most of the tugs in existence were small, under-powered craft that could handle only the old, animal-drawn barges. The handwriting was on the wall: adapt or die. An individual who owned but one or two barges was doomed. Buying a horse or mule was one thing, but to invest in a tugboat was simply out of the question for the majority. An analogy for today would be someone with a "Mom and Pop" grocery trying to move up to become a supermarket.

So far as the towing contracts are concerned, the first step was an inspection by the State of the boilers, engines, and the overall condition of one's tug. Tugs had to be able to tow six loaded boats (or approximately 1500 tons) at three miles per hour. Since many of the tugs had engines of only 70 to 140 horsepower, their ability to perform the work was at best marginal.

Going through the records, one finds some tugs that were rated by the inspection team as too small for the work were still awarded contracts. WM. F. FOX and LIBERTY are cases in point. Although they were tugs with no sleeping quarters, they were hired anyway; and there were other boat owners like Ben Cowles of Buffalo, who claimed his tugs were of a higher horsepower than the State inspectors found.

After being awarded contracts, owners had some substantial up-front costs. A deposit of $250 was required in case they failed to execute a contract. Next, there was a $1000 bond of faithful service and finally a $500 labor bond to guarantee payment of wages. In return, the State paid between $1125 to $1500 per month for each tug in 1917. By 1920, the rate was on a daily basis of $57.50 to $85.00. The difference was based on horsepower and condition of the tug. In addition, the State provided all the coal needed. Why they did this in light of the problems it generated is a mystery.

The wording in two sections of the contracts was the focal point of many disagreements between the State and the owners. One section had to be rewritten, and the other had to be strongly enforced. The contracts for 1917 stated the tug must be in a condition *ready* for service at all times 24 hours each day. To the owners, it said, "I have to be ready, but I sure don't have to *run* 24 hours per day." Some worked only 8 to 12 hours a day, which of course, had a disastrous effect on the movement of traffic. This was certainly not what the State had in mind when they drew up the contracts. A later change read, "If the tug fails or refuses to *operate* 24 hours each day, the contract will be terminated." I have correspondence to show the difficult time the State had over this issue, differences that led to bad feelings and threats by the State which said, in essence, "Shape up and run more hours or we'll give your contract to someone else!"

Another disagreement concerned which type of boats that had to be towed or assisted. Owners claimed they only had to tow or assist barges and scows – not steamers, yet the contracts clearly stated that they would tow boats of every description and character.

The frustration of finding and hiring tugs soon turned to aggravation when it came to operating them. There were the usual delays due to accidents and mechanical failures. Shippers were unhappy with the slow service. From time to time, there was a shortage of coal, and there were disagreements with coal companies over prices.

On September 4, 1920, for example, Fort Plain Coal Company delivered 11,840 pounds of coal to the tug P. C. RONAN at $16 a ton. Before the month was out, Edwin Walsh, Superintendent of Public Works, wanted an explanation of this high rate, since the average at the time was $9 per ton. Mr. Walsh either was not aware or had forgotten that coal then, like oil today, is generally bought on long-term contracts at a favorable price. When forced to buy coal on the open market, car by car (as Fort Plain did) the price is always considerably higher. Fort Plain Coal had plenty of records to back up their rate, and the Company offered to open its books to State Officials.

Another incident is a classic example of "Murphy's Law." Since the damage was not great, there is a humorous aspect to the story. Here is part of a letter written by Superintendent W. W. Wotherspoon to Mr. James W. Follette, owner of a damaged barge:

> The tug GEORGE E. LATTIMER was approaching Captain Tanner's boats to take a line and as the tug neared the boats, the captain of the tug intended to give his engineer two bells, the signal to reverse, in order to stop her speed. Upon pulling the first bell, the rope broke, making it impossible to ring the second time. The one bell being the signal to go ahead, the engineer started ahead instead of reversing and the tug rammed the canal boat "VIDETTO," breaking her planking. The captain of the tug admitted his sole responsibility for the accident and says his company intends to repair the damage for Captain Tanner.

The Superintendent of Public Works probably gave a cheer when he finally managed to stop State towing at the end of the 1920 season. It was now time for the canal to start functioning as it was designed to do.

Oil/gas motorship, circa 1920.

Fleet of steel barges, Lock 28A, Lyons, New York, 1919.

Federal Control

The United States declared war on Germany on April 6, 1917. It was the beginning of our entrance into World War I. At the time some Government agencies were surveying the condition and capacity of U.S. inland waterways to see if they could be used in the war effort. The State Engineer and the Superintendent of Public Works offered our unfinished canal, their services, and those of the State to hurry its completion as soon as possible. A committee from the National Council of Defense came, looked, and was not impressed. Aside from the fact that the canal system wasn't finished, new boats had not yet been built and the supply of old boats was inadequate.

This setback didn't dampen patriotic spirit, however. The State was so eager to help, a convention was held on August 1, 1917, to look for ways to complete the canal ahead of schedule and to wake Washington up to its value. The result was a petition to President Woodrow Wilson, the Secretaries of War and Commerce, and the Committee of Inland Waterways calling attention to the availability of the New York State canals and urging their use to the fullest extent. Although State officials were initially ignored, a tremendous push was made to finish the canal so that it would contribute to the war effort.

On January 31, 1918, State Engineer Frank M. Williams appeared before the Senate Committee on Commerce on a related matter, and this gave him an opportunity to push for the Federal Government to use the canal. He informed the committee that the canal would be opened throughout its entire length the next spring. But there would be a snag: the canal would be open, but there would be few boats to operate on it. The State Engineer bluntly offered them a solution: If the canal was to be utilized, then it was the duty of the United States Government to either build a fleet or assist by some method to provide one.

Two months passed before Washington made a move. On April 10, the Federal Inland Waterways Committee appointed by Mr. William G. McAdoo, the Director General of Railroads, asked the New York State Canal Board for cooperation of the State to help coordinate the use of the railroads and the canal system for the period of the war. On April 18, formal announcement was made by Mr. McAdoo that he would secure boats and establish an operating organization to utilize the State canals.

Joy rang through the State. Canal advocates were ecstatic that they had found an unexpected ally who could bring about in months what could take them years to accomplish. They were referring to a short cut to building up the canal traffic by diverting commerce from the railroads. Here was the supreme power, having absolute authority over all transportation, that could route traffic where it pleased. Since the railroads were choked, it seemed inevitable that the canal would get a large share of the traffic.

It is difficult to believe those in authority could have been so naïve. Here was a classic case of the fox guarding the henhouse. We have the Director General of Railroads – the sworn enemy of all water-borne transportation – in charge. It boggles the mind to think a layman would expect him to help competing system, yet men with expertise in water transportation did just that.

Disillusion came with lightning speed with the announcement that canal and rail rates would be equal. After some pressure, this ruling was changed and canal rates had a twenty percent differential. As further events would prove, the change was nothing more than throwing the canal folks a bone. The next announcement really caused an uproar across the State: No private lines were to be allowed on the canal. The protest screams were heard in Washington and Mr. McAdoo magnanimously disavowed any intent to forbid private operation of boats.

All the canal advocates' suspicions were justified when the railroads were turned back to their private owners after the war and the canal remained in-the control of the Federal Government. To add insult to injury, the Government bore some of the expense of maintaining the railroads during the war, but not one penny was ever given to the State of New York toward building, maintaining, or operating the State canal systems.

During the war, there was one last attempt to change the way the Government ran the canal. After some four months of trying, an interview with the Director General was granted for October 25, 1918. Mr. McAdoo was completely pro-railroad. He finally closed the interview by taking the intimidating position that as long as *he* had the power, no freight would go by canal as long as the railroads could carry it.

A statement made in 1919 shows the bitterness toward the continued control of the canal by the Federal Government. A paper read before the State Waterways Association by Edward T. Cushing of the New York Produce Exchange says, in part, "It is for the interest of the Government to kill any competition of the canal with the railroads, for even if the railroads were returned to their owners, the Government would still guarantee their earnings. No sane man will compete with the Federal Government. What an object lesson in paternalism! The fear of it today is paralyzing the operation of the greatest inland waterway in the world. Here is the biggest thing ever played in railroad history. The stake – one hundred and fifty million dollars of the people's money invested in the Barge Canal; the contestants, the United Railroads, backed by the Federal Government, against the people of New York."

Early in 1920, the railroads were returned by an act of Congress to their former private owners. The act did not include the Barge Canal. Even in the 1920's, Government agencies were adept at stonewalling. The people continued to be suspicious of the railroad fraternity, but nothing could be proven. Protests failed, and because of the importance to the whole country of the chief features of the bill, it of course could not be delayed to change one "canal item." The bill passed and with it, control of the canal shifted to the Secretary of War. New York State officials spent a considerable amount of effort persuading him to give control of the canal back to the State. Success was realized just before the 1921 season opened.

How well did the Government do in the years it operated the canal? The report of the Director of Inland Waterways of the United States Railroad Administration for the year 1918 attributes their failure to the fact that equipment was obsolete and inadequate, and time too short to get an operating organization up to speed. That is doubtless a fair statement, for it does appear that they did as well as they could. The report for 1919 shows a loss of $506,807.38 and the failure was admitted. The Federal Government had a different excuse this time. The unsuccessful operation now was blamed on the fact that modern power units which the Government had contracted for had not been delivered; furthermore, they said that tow boats available for the movement of the new steel and concrete barges that had been delivered were inadequate.

Throughout 1920, the Government had its full complement of floating equipment with a price tag of approximately $4.5 million. They had twenty steel steamers with a cargo capacity of 350 tons, 51 steel barges with a capacity of 650 tons each, 21 concrete barges with a capacity of 520 tons each, and three wood barges with a like capacity. Although the fleet had a total capacity of some 600,000 tons, it only carried 197,017 tons, less than a third of its capacity. The so-called "old and obsolete" boats carried over one million tons with much greater efficiency. The average time per trip for the Government's boats (Buffalo to New York) was 14.1 days, as against 7.9 days made by the independents' boats. It's no surprise, for what incentive could a Government boatman have to perform well? He worked for a non-profit organization, who, in addition could not care less how well he did his job.

33

Conditions got so bad that shippers of flax seed from New York switched from Government boats to an independent operator. Immediately, the Government decreased its rate on flax seed; and since the former rate was fair and reasonable, many doubted the new one could be profitable. State Superintendent of Public Works Edward S. Walsh was furious over what he called "typical destructive competitive methods" used by the Government. While no one can fully condone the Government's action, the practice certainly wasn't theirs exclusively. Businessmen have been trying to bankrupt their competitors since the first caveman slashed the price of his war clubs and spears.

One facet of Government operations, however, Superintendent Walsh had a good reason to yell about. "The utter incompetency and rank carelessness of Government employees manning the boats placed the canal structures in constant jeopardy," he said. "Navigating the waterway with complete disregard of rules and regulations the Government boats wrought havoc with the channel buoy lights, badly damaged locks, time and again, were in collision frequently with other craft, were sunk here and there in the channel, and in one instance, almost completely destroyed a bridge." He also spoke of reports of these same men being too drunk to operate the boats safely. Had this happened with a private company, he noted, the drunken boatmen would be barred from the canal.

A word should perhaps be said in defense of the men who worked on the Government boats. Dick Garrity, who wrote *Canal Boatman* (published in 1977), and Captain Archie Thurston (who was a boatman from Montezuma, New York), both had experience on these boats and recounted some of their horror stories. Because of the design, the power units were very difficult to steer. To complicate matters, the steel barges were almost impossible to tow in a straight line and therefore they did indeed cause a lot of damage. Boatmen were twice cursed when towing the experimental concrete barges, for it took very little to stove a hole in them. They were such a fiasco that the Munson Steamship Line that bought the fleet after the war refused to include the concrete barges in the deal. At any rate, that sale brought to a close a sorry chapter in the history of the Barge Canal.

The First Years 1918-1925

Reading about Federal control leaves one with the impression that little if any private towing was done on the Barge Canal system until after the canal was turned back to the State in 1921. Not so. It was almost exclusively on the Erie branch where the bulk of Government-control shipping took place. Overall traffic on the system was light for the 1918 season, but independent shippers carried nearly all the cargo on the Champlain, Oswego, Cayuga-Seneca, and Black River branches.

Although few private companies were active during this period, some are worth mentioning. The oldest and largest was the Lake Champlain Transportation Company of Whitehall, New York, usually referred to as "The Line." This company was active from the very beginning of the original Erie Canal and remained in business until the mid-1930's. The bulk of their business was always on the Champlain Branch.

A pioneer in the movement of petroleum was Standard Oil Company of New York (Socony). With one tug and one barge, Socony moved 14,000 tons in 1918 to Schenectady, Little Falls, and Rome on the Erie Branch and to Fort Edward and Whitehall on the Champlain Branch. One company appears to be the first manufacturing concern to operate its own boats, the now-giant General Electric Company. They moved finished products from Schenectady to New York City and returned with raw materials, running three deck-loading barges during the 1918 season. Later, they added a tug of their own.

Another large but short-lived company was the Ore Carrying Company of New York, which operated 22 barges carrying ore from Port Henry on Lake Champlain to New York City, with coal on the return trip.

One of the earliest users of the canal was Shippers Navigation of Syracuse. Although this firm was incorporated too late in 1916 to take advantage of the new canal, it was in full swing for the 1917 season, only to be taken over by the Federal Government in 1918. Shipper's Navigation resumed control in 1919 with nine cargo steamers and 50 barges of the old Erie canal type – 98 feet long by 17 feet wide and having a capacity of 250 tons.

Although gone from the canal for decades, Transmarine Corporation, which began in 1920, has left a lasting monument. They were responsible for the creation of the Port Newark terminal on Newark Bay, a terminal that is still in use.

Although total tonnage was less in 1921 than 1920, with Federal control off, more companies were beginning to use the canal, with two firsts credited for 1921. One was the entrance of the Sugar Products Company of New York with a four-tank barge (158 feet long) carrying molasses in bulk and having a pump supplied by its own steam boiler to off load its cargo. In three trips, it moved 190,000 gallons for a rate less than half the rail rate. The other first was a new type of vessel never seen before on the canal. Interwaterways Lines of New York had five twin-screwed steel cargo boats built in Duluth, Minnesota, to carry dry bulk cargo such as wheat, flax seed and pig iron. They carried two 140-horsepower semi-diesel engines on boats that were 242 feet by 36 feet. After a few trips, the company experimented to see if these new boats were powerful enough to also tow a couple of barges. It worked, so the company could tow two 500-ton barges on each trip.

Total tonnage on the canals increased an impressive 67% for 1922 over 1921, and this seemed to signal that the Barge Canal might well be coming into its own, with its earlier difficulties behind it.

Seven new companies entered the field in 1922, one of which was headed by William E. Hedger of New York City. Hedger is remembered for two things: for being the unofficial Sulphur King and for his "boxes." His boxes were formerly harbor barges, square-ended and larger by 10 feet in width than most canal barges. Unfortunately, Hedger is also remembered for the debts he left behind when he went broke in the mid-1930's. His company was dissolved by State proclamation in 1937. An innovation by the Inland Marine Corporation in 1922 proved to be noteworthy. To increase production of grain shipments from Buffalo to New York, Inland Marine installed a floating elevator at Albany, where grain could be shifted from canal to river barges that could carry 33,000 bushels – three times the capacity of canal barges! Inland Marine also received an added bonus by not having to take canal barges to New York; their turnaround time was much less than that of their competitors.

From 1918 through 1925, growth was not impressive, averaging 169,000 tons per year. This poor showing triggered a report in 1926 from Superintendent of Public Works Fredrick Greene. It must have sent shock waves through the canal advocate community, and average citizens must have felt betrayed once again by the politicians. In effect, the report said the canal was a real loser. Mr. Greene's opening sentence was a masterpiece of understatement: he said the canal traffic "had fallen short of expectations." The annual capacity has always been stated as 20 million tons, but in 1925, only 1,238,844 tons were moved. The real eye opener came in his next paragraph. It would have been cheaper, he said, for the State if all the freight carried on the canal were put in railroad cars and the State pay the freight. The railroads must have had a field day with that! Mr. Greene said a fair rate by rail from Buffalo to New York was $3.70 a ton, whereas it had cost the state $4.51 a ton for all the freight floated on the canal.

To a counter argument to the bleak picture, Mr. Greene responded in this fashion:

> It has testified that the canal saves the people of the State 50 million dollars annually in 'depressed' rail rates. This has not been proved to my satisfaction. The old Erie Canal undoubtedly served to 'depress' rail rates; this however, was before the existence of the two rate-regulating authorities: The Interstate Commerce and the Public Service Commissions. Having these regulatory bodies, the questions naturally arise:
>
> 1. Would these authorities have allowed rail rates to be increased 50 million dollars a year, if canals were not built?
>
> 2. Are states lacking canals overcharged by the railroads 50 million dollars a year or in proportion according to the amount of freight carried?
>
> 3. Is not a club, costing 10½ million dollars a year, an expensive weapon to hold over the heads of the railroads?

The key question is: Why the dismal showing? While there might be a variety of reasons, it all really boils down to one: "Ice." The canal is closed down five months each year because of ice. Mr. Greene also felt the clearance of only 15 feet on the immovable bridges was a limiting factor to canal growth. With excellent hindsight, we can now say that bridge height was not a major factor in canal traffic growth, for this could have been offset by continued enlargement of the canal prism and the locks, as Canada has done with the Welland Canal. An even better example is the St. Lawrence Seaway, which is ice-bound all winter, yet has proved to be very successful.

The original Erie Canal had succeeded in spite of ice and low bridges. The investment in a mule-towed canal boat was so small, and winter carrying charges so little in comparison with the business done, that Erie boaters could afford to remain idle during the closed months, and the low bridges didn't hamper them at all. Financial matters are vastly different, however, when a modern, self-propelled vessel, costing from $100,000 to $250,000, must lay idle five months while heavy interest charges and insurance costs accrue.

Another area of great disappointment was the use – or, more accurately, non-use – of canal terminals and the Gowanus grain elevator. Mr. Greene's report addresses this situation like this:

In the past, attempts to increase canal tonnage have been confined largely to building more or less expensive structures along the Canal and at points off the Canal proper. In 1911, the people approved a bond issue for this purpose, amounting to $19,800,000. Immediately thereafter, different communities began scrambling for wharfs and terminal buildings. The record in this office of the freight predicted – all predictions solemnly affirmed by local Chamber of Commerce, and other organizations – when compared with the actual business done at these terminals, is enlightening and at the same time disheartening.

In connection with the Barge Canal System, the State has 66 terminals (piers or bulkheads) at which there are 53 warehouses; these terminals range in cost from a few thousand to more than a million and one half dollars. During the past two years, no freight was handled at 49 of these terminals and only five warehouses were used for canal freight.

The following terminals have received *no* canal freight either in the warehouses or out of them during the past two years:

Location	Cost	Location	Cost
Rouse's Point	$ 55,069	Amsterdam	$ 94,622
Plattsburg	159,000	Little Galls	87,095
Port Henry	90,346	Herkimer	75,363
Whitehall	87,646	Ithaca	63,201
Schuylerville	78,312	Lockport	76,402
Cohoes	95,438	Fonda	70,003
Flushing	407,172	Hallett's Cove	606,495
Mott Haven	1,039,038	Greenpoint	1,608,999

Official records show that many of our terminals have never been used for canal tonnage since the day they were built. The terminal at so important a city as Albany, costing $312,914, has received but one small canal shipment (lumber) during the past two seasons and in that time no canal freight at all has gone in or out of the freight house.

The early years of the Barge Canal, as might be expected in any new venture and as contemporary New Yorkers – both canal promoters and canal opponents – came to realize, were filled with birth pains and uncertainties. The fact that the United States, and hence also the new canal system, almost immediately became involved in the war effort didn't help things. Neither did control during the war years by the Federal Government. The years from 1918 to 1925 were uneven ones at best, even after the canal was turned back to the State in 1921. At any rate, total tonnage for 1925 was 2,344,100 tons vs. 2,602,000 tons twelve years earlier when it was formed. If total tonnage is a measure of success, at the close of 1925, the Barge Canal was not quite where it was in 1913. It's time now to leave the history of Barge Canal and look at life in a tugboat in 1977.

CHAPTER 3

THEN AND NOW

The first active company I researched was the one owned by Captain Marty Kehoe of Latham, New York. This led to the first of many trips on one of his tugs. I wanted to make comparisons between living on the MATTON 10 in 1941 and life on a tug in 1977. [All my trips were taken on the Champlain Branch of the Barge Canal System.]

After the first of many interviews with Marty, I asked his permission to take a trip on one of his tugs. At that time, he owned five, some of which would not be suitable for me to travel on because of size or crew temperament. I wanted a tug with a mix of old timers and new men, a crew that would tolerate having me underfoot, asking questions and snapping pictures, plus a tug large enough so that I would not crowd the crew. His choice of the ERIN KEHOE was perfect in all respects. Over the years, I have met and enjoyed traveling with over thirty-six Kehoe men plus many bargemen. One, Bob Gordon, became a great friend.

To catch up to a tug is not the same as grabbing a ride on a bus or train. It's impossible for tugs to run on a tight schedule. They are affected by weather, traffic, breakdowns, and terminal delays. After this first trip, I would call Marty and get a rough estimate as to when the ERIN would be at Lock 1 on the Champlain Branch of the canal. I chose this lock because it was close to my home and seemed a safe place to leave my car. The lock tenders were very good about watching everyone's cars. I would leave home a day early to make sure I didn't miss the connection. The crews often have a more difficult time because they have a specific crew change day. If they are lucky, the tug will be near their home when the day comes; if not, it's a real headache to catch the boat. This is especially true for the men who live in Florida or other distant states. The airplane, of course, has made it possible for these men to live that far away and still work on the tugs.

That was the first difference I noticed from my time on the MATTON 10 in 1941. In those days, the crew men lived on or near the canal in places like Waterford, Whitehall, Syracuse, Buffalo, and Tonawanda. It surprised me that the ERIN KEHOE'S engineer, Jack Vetter, lived in Florida and yet worked on the upstate New York canal. To me, one of the strangest of all was bargeman Harry Tungate, who lives in the tiny town of Dry Ridge, Kentucky. To my mind, canal boatmen just don't live in places like *Dry*

Ridge! Harry was special in one other way. He was second to deckhand Tommy Davis when it came to telling tall stories. Getting from Dry Ridge, Kentucky, to the village of Keesville in upstate New York is no small task. On this, my first trip, he had to do just that. Unfortunately for Harry, he arrived late at night. With his rugged build and heavy beard, he is a formidable sight. Since the tow had not arrived, he had to hang around town. Well, many small towns are wary of strangers, especially late at night, and so the local constables were going to lock Harry up as a vagrant. It took a little talking on his part before he convinced them who he was and what he was doing in Keesville.

I was very nervous about meeting the crew. I had hoped that Marty Kehoe himself would volunteer to take me to the lock and introduce me to the crew. I finally got my courage up and asked him if he would pick me up at my motel in Latham and take me to the boat. He was very nice about it and admitted that most nights he was up late anyway. Arriving at Lock 1, we had about 15 minutes to wait and my worry meter was running at full tilt. What if the crew didn't accept me?

First to stick out his hand and say hello was Gordon Blood, the Mate who had just come on midnight duty. I learned later he was moonlighting for Kehoe from his regular job with the Bushey organization of Brooklyn. This moonlighting is quite common on the canal today, because there is a constant shortage of qualified wheelhouse personnel. Some like Gordon Blood and Amos Yell will work for other companies on their time off. This is a lifesaver for small outfits like Kehoe, who otherwise would have to run shorthanded or tie a tug up.

Marty said his good-byes and left. Gordon led me to the galley, where he introduced me to the other two men on the midnight-to-six a.m. watch. After a few minutes they, too, made me feel at home. Gordon then took me below, got me linens, blankets, and showed me where to bunk. I undressed, crawled in and was dead to the world in less than two seconds.

I was up at 5:30, hungry and anxious to meet the rest of the crew. Charlie Passero, the cook, had been up since 4 a.m. Breakfast held no surprises for me. It had not changed since my time on the MATTON 10 – several kinds of juice, oatmeal (always oatmeal!), eggs, bacon, pancakes, and home fries. At this rate, I would gain a ton on the trip. Protocol and custom are just as important on a beat-up old working tug as it is on a spit-and-polish Navy vessel. Example: Passero saved me from committing a *faux pas* by showing me where to sit, and at each meal whether I should eat at the first or second setting. There are not enough places for everyone to eat at the same time, so the rule is that the oncoming watch always eats first.

41

In explaining earlier how the watches worked, I failed to mention the cook. Although he works 12 or more hours, he doesn't work six on and six off. Starting at four a.m. till seven p.m., he will try to get a nap in the afternoon. And like the housewife, his work is never done. Even when we are tied up for fog, he still has to cook while the rest of the crew can rest. One thing is a little different on the ERIN KEHOE. Captain Bob Gordon, being the kind of man he is, does not always follow the hard-and-fast rule that all the senior men get the more desirable early watch. He allows the men to switch off so everyone gets a fair share of the early watch. He was also flexible if some of the men wanted to work the normal 14 days on and 14 days off, or seven on and seven off (if the man's counterpart agreed to work that way.)

My first trip was not too well organized, I have to admit. What with a tape recorder, notebooks, cameras, and extra lenses, I was in worse shape than the proverbial one-armed paper hanger. Since I didn't know the routine of the tug, I couldn't plan things very well. Trying to decide when to tape an interview or grab a photo opportunity caused me to look anything but professional. I soon learned that tape recording was a big mistake, because the normal engine noises, vibrations of various bits and pieces of the boat, and other people talking drowned out the person I was trying to interview. After this trip, I decided I would use a notebook instead of a tape recorder.

The prominent feature of canal travel, of course, is the locks. Entering each one presents a different degree of difficulty. This varies with the wind and the amount of water flowing over the dams. When talking about the water flow, the crews always refer to the number of gates that were open in the dam. When the water is low, all gates will be closed. As water rises and flows over the dam, the amount is controlled by opening gates (you might think of them as faucets or valves) one by one. During a heavy spring run off, all the gates will be open, yet there still will be a hell of a lot of water rushing over the dam. This creates fast currents and strong eddies that test a boatman's skill. In many sections of the canal, high water can cause the tows to be tied up because they can't get under the bridges – hence the name of this book. This condition can go on for many days. To appreciate the problem of entering the locks under inappropriate circumstances, you must look at the size of the tow. They are almost 300 ft long and 43 ft wide. The locks are 300 ft by 45 ft, which leaves very little margin for error and no room to maneuver.

Besides being aware of the wind and flow of water, the man in the wheelhouse has to contend at times with blinding rain, fog, snow, and pitch black nights and, upon occasion, a deckhand who is not too swift in handling lines.

His first challenge comes with "shaping up" the tow. This means lining up the tow so he has a straight-in shot at the lock. At some locations, he has a long straight line to the entrance, but at others, he comes around a bend with a very short distance to line up the tow. About the time the tow is lined up, the deckhand on watch and the bargeman on watch go to the forward end of the barge, one on each side and with hand signals help with the final alignment.

Barges do not carry permanent rope or rubber fenders as tugs do, but something is needed to avoid heavy scraping of the barge against the concrete walls of the locks. This is accomplished by "walking the dog." Let me explain. Each man has a small rope fender tied to a short line. They drop it down between the lock wall and the side of the barge and start walking forward. After about 15 or 20 feet, they pull it out, walk back and drop it over the side again. After about one-third of the barge is in the lock, they can stop. The tow will now stay in line and is controlled by the throttle alone. It really does look like someone walking a dog on a leash. The two men now get ready to put the lines out to hold the barge in place in the lock. These lines will have to control both forward and back-ward motion of the tow.

The tow is now crawling very, very slowly toward the closed gates. Even at this dead slow speed, if the tow hits those gates, it's "good-bye Gertie, that's all she wrote." It does happen now and again. In fact, this tug did it in the Federal Lock at Troy, New York a couple of years ago. Both men heave their lines around a bollard (steel post) on the lock wall and take a few turns on the barge cleats. [Steel hardware with projecting ends to tie hawsers to.] As the tow inches forward, the lines slip through the cleats. By adding or taking off loops, they can ease the tow to its final stopping place. A definite skill is required for this operation. I have seen experienced deckhands misjudge the speed of the tow, snub down too quickly and part a line as if it were a piece of thread. A snapped line can do great damage to one's body.

The man in the wheelhouse does not rely entirely on depth perception to gauge how close he is to the gates. Some have a spot on the wall they use as a mark to stop the tow. Without an obvious signal, he will throw the engine in reverse for a few seconds to stop all forward motion, as the men on the barge take turns on the cleats. The job is not quite finished.

As the tow moves into a lock, there is not enough space on each side for all the water to flow around the barge; therefore, it acts like a monstrous plow and shoves a wall of water in front of it. As a result, as the barge gets very close to the gates, the water pushes the tow backward. Again, the deckhands control the movement with the lines.

The time to fill or empty the lock gives the Captain or Mate a 10-to 20-minute breather, to stretch his legs or maybe walk back to the galley for a cup of coffee. He may also use the time to call the office. All companies like to know where their boats are each day. During this time, the deckhands are still at their posts, because now they must control the tow in different directions, up and down.

There are eleven locks on the Champlain Branch; although they are numbered 1 through 12, number 10 was never built. Locks 9 and 12 are special. Route 22 runs alongside Lock 9 with a pull-off area for the tourists. I always enjoyed watching the crews when we come to this lock. They protest they don't like to be gawked at, but most of them seem to wander out on deck under some pretext or other.

Lock 12 is special for other reasons. It is not only the northern terminus for this section of the canal, it's the only lock that runs through a city: Whitehall. Whitehall is the home of many of today's boatmen, as it was in the days of the old Erie Canal. Men from this tug – The Gordon Bros., Amos Yell, Dave Oliver and Jimmy Day live here. The boats have their laundry done here, and many times they grub up [buy groceries], too. It also is the birthplace of the U.S. Navy. It was here that Benedict Arnold built the makeshift fleet which defeated the British in the battle of Valcour Island in Lake Champlain. There is a small museum in town that is well worth one's effort to visit. And finally, it was the home of one of the largest and most successful boat businesses that operated on Lake Champlain, or the Erie and the Barge Canal systems. It was the Lake Champlain Transportation Company, more affectionately referred to as the "Line." Whitehall today pays little attention to the comings and goings of the boats, except for the families of the crews. They frequently meet the boats to bring a missed piece of clothing, to discuss a family problem or just to visit a few minutes with their loved ones.

Technically, when you leave Lock 12, you enter Lake Champlain but it certainly doesn't look much like a lake. The first 11 miles are through a narrow twisting waterway aptly called the marshes, at times barely with enough water to navigate. Passing another tow can be a little nerve-wracking and sometimes the men lose their night vision with all the fishermen on shore with their blinding gas lanterns. On those occasions, I have heard a few choice words flung around the wheelhouse.

The lake really begins at Stony Point and the trip from there to the oil terminal should be described using poetry. Running between the Green Mountains of Vermont on one side and the Adirondacks of New York on the other, is both scenic and restful. I close my eyes and see the birchbark canoes of long gone Indians and question if we should be disturbing the peaceful tranquility of this beautiful body of water. But we are, and we carry much needed fuel oil to the towns on both sides, plus jet fuel to keep the planes flying out of Plattsburg Air Force Base.

As we neared Port Douglas [on the west side of Lake Champlain across from Burlington, Vermont], I expected to see a typical concrete pier, which we would pull alongside to unload our cargo. Another assumption of mine that proved wrong. The terminal consisted of a pipe line strung on a steel catwalk out into the lake and ending with a tiny platform sitting on a dolphin, where the hoses can be hooked up. It doesn't take much imagination to appreciate what it is like to battle wind and waves in this exposed position to maneuver a tow alongside the dolphins. Now try it at night with a swirling snow storm blowing and you will see real seamanship that many folks think is required for only deep sea vessels. Nowadays, there is another fear riding with these crews, the nightmare of pollution and the possibility of stoving in a barge. Pollution is a deep concern of all boatmen and often the reason many deckhands no longer aspire to become Mates and Captains, and become responsible for such accidents.

All the men in the wheelhouse pride themselves on landing their tow on the first try. In addition to weather, there is the human factor that can foil their landing. An example: The water is shallow around these oil terminals, and there is little room for error. On this trip, the bargemen started taking on ballast for the trip back before the oil was unloaded. Captain Amos Yell grounded the barge and had a hell of a time bringing her into the dolphins. He was furious and rightly so. To add insult to injury, the bargemen had the raw nerve to complain because Amos was taking so long to land the tow. By taking on ballast at this time, he could get a little more sleep the next day while we were running down the lake, the time when we would normally take on ballast.

After tie up is completed, the tug is separated from the barge and moved out of the way. This is also done during loading. This is done for three reasons: for safety when handling volatile fluids, because of the extreme changes of angle and depth the barge goes through during load/unload

45

and finally it gives the tug crew time to refuel, grub up, and take on fresh water. Depending on the viscosity and outside temperature, it can take from 6 to over 24 hours to pump out a barge. One other factor affects pumping time, and that is how much head they are pumping against. For example, when off-loading at Port Douglas, the fuel has to be pumped up a sizable hill to the holding tanks. When loading and unloading the tables are turned. The bargemen are very busy while the tug crew can relax.

As soon as the barge is empty, it's time to go. These boats work 24 hours a day. While going down the lake, water ballast is taken on. During this time, the tug and barge are connected by safety lines only. The cables are not put on until all the ballast is in, otherwise the deckhands would be constantly adjusting the steamboat ratchets (used to tighten the cables) to match the draft of the barge. Ballast is necessary for two reasons – one, to clear the bridges, and the barge has to be low enough in the water for the man in the wheelhouse to see over it. Some barges, such as Carl Ecklof's E-17 (formerly TEXACO 798) are a real pain because the crew's quarters and pump house are built on top of the deck, where in most cases they are built into the barge.

Standing on the deck, with my hands wrapped around a steaming cup of coffee and breathing the crisp, clean air, stimulates my soul. It feels so good to be alive, and suddenly, I feel so sorry for all those poor folk on shore who can't share this moment with me. The world and all its problems are a planet's distance away, and God! I wish I could freeze this moment in three dimensions.

But we settle into normal routine. Breakfast is over and the off-watch has hit the sack. The Chief Engineer has gone below to check the engines, clean strainers, and maybe do some painting. Low water causes the tow to stir up mud from the bottom. This muddy water carries leaves and junk that can quickly plug the strainers preventing the engines from getting sufficient water to cool them. If it's a fair day, the deckhand will wash down the boat, wash the wheelhouse windows and polish brass. And of course there is always painting to be done. The standard joke is how they would make great housewives because of their cleaning skills. This brings to mind something that happened on another trip I was on. Two of the most imposing deckhands I ever met were Tommy Davis and Larry Pauquette. In a brawl, you would definitely want them on your side. One day this dynamic duo were in a heated argument about the merits of one particular detergent over another one. Standing nose to nose, they sounded like a couple of jaded housefraus. Naturally, I had to have a photo of this

Larry and Tommy argue the merits of cleaning products for benefit of the author.

for my book. A couple of years later, someone told me they had set me up for this act. The joke was on me of course, but I loved it. It made me feel accepted as part of the crew because they were dreaming up practical jokes on each other all the time. Sometimes the normal routine is interrupted for the deckhand because the Captain wants a cup of coffee or a little relief at the wheel.

Except for the constant checks on the engines, most work is not on a rigid schedule as it is in the Navy. As long as the jobs get done, the Captain or Mate do not direct the work in minute detail. On occasion, I have heard them remind a cook or deckhands of some overlooked or put off chore. Life is a little different for the bargemen. They don't have a separate deck and engineering department. Since there are only two men they do it all – cook, clean, operate and maintain all the machinery, and also act as deckhands. Theirs is a highly skilled job.

If all this maintenance wasn't done, these tugs wouldn't last the forty to fifty years that most do. The old timers will tell you that the crews worked much better in the 1930's and 1940's and there may be some truth to that, but I suspect that each reigning generation feels that way about the previous one. On the other hand, a couple of the younger men, Dave Kehoe (deckhand who is related to Marty Kehoe) and Jerry Kase (engineer) really impressed me and I think they could hold their own with anyone from the "old days." How a man handles his tug or tends his engines can make a real difference in the profit and loss statement of owners like Marty Kehoe. Men like Bob and Paul Gordon, Dave Oliver, and Amos Yell treat a tug as if it were their own.

After leaving Lock 1, I looked forward to reaching the 125th Street bridge at Waterford. Living next to a river, one of my reference points became bridges. My world ran between the bridge at 125th Street and the one at 112th Street, long hours spent leaning over the rail, watching and dreaming as the tows moved below me. Warm, lazy summer afternoons, lying on the bank staring across at Matton's Boatyard. On a lucky day, they would be floating down their drydock to put a tug or barge on for repairs. Our summers revolved around the river. We fished in it, swam in it, (never gave a thought to pollution in those days), but mostly we watched the boats.

In the mid-1930's, at the start of the horse racing season at Saratoga, a white-hulled-beauty would come up river and tie up to the wall at 122nd Street. It was a big event for us kids. To us, that boat was the biggest, most beautiful yacht in the world, with all its shiny brass, polished teak and mahogany decks. It belonged to a man named Page, who someone told us had invented the bicycle coaster brake. He did own the New Departure Company that manufactured them. In any event, in the middle of the great Depression, to us, he had to be the world's richest man. Their chauffeur had driven their limousine ahead of them and was waiting for their arrival. Each day he would drive them to the races and bring them back each evening. This event was what we waited for each day. While the yacht was beautiful, the special attraction was the crew. They all wore white Navy-type uniforms. The Captain was dressed in blue and stood at the head of the gangway as the crew formed two rows as side boys do in the real Navy. The only thing missing was a boatswain piping the owners over the side. Even today, I can hardly believe they went through this ritual twice each day.

I doubt if you could find a crew today that would go through this pompous charade. But, you must remember, this was the great depression and jobs were almost impossible to find. What was imprinted on my mind was the fact that they had a terrific job, when my Dad had no job.

Farther downstream, from Troy to Albany, are the sights best passed over: grimy, hollow shells of factories, failed businesses of every kind. Very little came back after the Depression. This scene repeats itself all the way to the Battery. Yet the Hudson does have some spectacular sights to offer. As the river widens below Albany, even the city shorelines don't offend so much, because distance blurs the details.

At Albany, we stopped at the Mobil dock for fuel and water. With luck, there will be one or two other tugs there. Maybe a Bushey or Morania boat and then the crews get a chance to visit with old friends. It would be rare, indeed, if they didn't know someone on the other tugs. As the traffic becomes thinner each year, it's not too difficult to know who everybody is. Although they don't like to admit it, or even think much about it, all but the most dense see the beginning of the end of this way of life. This trip was a special treat for Captain Amos Yell. The FREDERICK BOUCHARD lay along-side us waiting their turn to get fuel. As a long-time employee of Bouchard, he found the Captain of the FREDERICK was an old friend. This was the closest I've ever been to a tug that overwhelms one by its size, power and beauty. This is a tug that normally works "outside" (a boatman's term for an "ocean going tug.") It does not have any up and down wheelhouse, such as the canal tugs, to see over empty oil barges. They never encounter low bridges, so it's cheaper and simpler to place a small auxiliary wheelhouse on a permanent tower that sometimes is 40 feet off the water. After introductions, the Captain asked me if I'd like to see the top wheelhouse. I was halfway up before his questions had left his lips. The view was awesome. I felt as if I were in the crow's nest of an old sailing ship. All the controls are at your fingertips and it was air conditioned, as was the rest of the tug. I was a little insecure up there as I tried to imagine what it must be like in rough weather. On my way back to the ERIN KEHOE, I couldn't help but think how palatial the other crews' quarters were.

Continuing down the river, I saw how light the traffic was. Sometimes between Albany and the George Washington Bridge we only see a half dozen tows or ships. One of the more unusual tows is the pushboat ROCKLAND COUNTY (owned by Bushey), a push boat on the river and about as close as we get to the huge multiple barge tows seen on the Western rivers.

The most dramatic change in the number of vessels is between the Battery and mid-Manhattan. In 1941, I lived for a time on West 30th remember when the river was filled with railroad floats and ferries running to New Jersey. There were also the majestic passenger liners, freighters large and small, and tugs, barges, and scows far too numerous to count. The airplane did the liners in and the evolution of the container ships moved all the cargo action to New Jersey, where the large areas needed for load/unload and marshalling were available. Most of the piers are gone now. With few exceptions, what is left is a sorry sight of collapsed rusty shells. But there is one place where there may be more activity than ever and that is in the "kills."

The name *kill* comes from to the Dutch, meaning "stream" or "channel" – the Arthur Kill and Kill Van Kull, e.g., separate Staten Island from New Jersey. As we hang a right at the Statue of Liberty and head into the Kills, life becomes a little tense for the man in the wheelhouse. We start to compete for space and try to avoid being run down by the container ships and the oil tankers that dwarf us. This part of the trip, while fascinating, also gives me the willies. This is the epitome of the place looking for an accident to happen. We are surrounded by and intimidated by acres of tank farms that line both sides of the Kills. There are high octane gas, oil, and every volatile, exotic chemical known to man. A fire here could level Staten Island and half the state of New Jersey.

On this trip we were followed and overtaken by a small launch. Easing alongside, the launch did not stop but kept pace with us, and a man in street clothes climbed aboard our boat. "My God!" I thought, "A water-borne mugger." Well, not quite. I soon learned that the visitor was a union delegate collecting dues. He certainly had a captive audience, for there was no place to run and hide, so the men might as well pay up. Also, there was another side to this union visit. It gave them an opportunity to air any grievance or just to catch up on union news and what was happening in the other boat companies.

Two features along the Kills are worth mentioning. One is the largest garbage dump I guess I've ever seen. The dump is mountain high, with a fragrance to match and at least a million screaming divebombing seagulls swooping down for a breakfast morsel. The other feature is a junk yard that lies in the water: a "marine salvage" operation of the highest order. I will-admit that there is an impressive variety of stuff: Staten Island ferries, tugs, barges, lighters, assorted small tankers, and some Navy craft.

As we continue steaming up the channel, I notice another change from the old days and that is the use of the radio. Instead of using the whistle, they use the radio for security calls. These are calls to warn other boats that we are approaching a bend, pulling away from a dock, or backing into a channel. Whistles may be more exciting and colorful, but crew members in their bunks don't need the extra noise. Still, old ways die hard. For example, when about to pass another boat, they will call on the radio and tell the other Captain they will be passing on one whistle. Incidentally, when on the radio, everyone is called "Cap." Just plain courtesy, because you can't tell rank over the radio. There is one whistle that is used and it's called the peanut whistle. It has a shrill high pitched tone that doesn't carry far, and the whistle is used by the men in the wheelhouse to summon a deckhand. In addition, the tug has a loud speaker system to direct deckhands and to yell at dumbbells like me when they get in the way.

O'Malley the historian at this point became O'Malley the tourist. I lolled in the lap of leisure, took some photos, ate too much, and thoroughly enjoyed myself.

CHAPTER 4

ON THE ERIN KEHOE – 1978

Paul Gordon was waiting with me for the ERIN KEHOE to enter Lock 1 on the Champlain Branch. He was a chief engineer and brother to Bob Gordon, captain of the ERIN KEHOE. I had never met the mate on watch bringing the ERIN into Lock 1, so Paul suggested that we put him on a bit. Paul would introduce me as a union inspector who was going to take a check ride with him. I'm not sure the mate believed us, but it turned out to be a good ice breaker. His name was Jimmy Day, and he hailed from Whitehall, New York. Even in the shrinking world of towing on the canal today, there are between 40 and 50 boatmen who live in Whitehall.

Jimmy Day was 32, tall, overweight, and in love with his job and food, but I'm not sure in what order. He began as a fireman on the railroad, spent three years in the Army, and went to work for Marty Kehoe in 1970 when the railroad abolished his job. In five years, he had his mate's license. Jimmy was an outgoing, noisy person, but he really did shine when someone was around to watch him steer. For a relatively new mate, he appeared to be a good boatman. I was impressed with the way he eased the tow into the locks. More important was his skill running in fog and at night on the Hudson River. There was one area in which he didn't seem comfortable, and that was working the different kinds of tows in New York harbor. If he had a shortcoming, it was his lack of interest in the paperwork that goes with the job.

I next met deckhand Tommy Davis. He could have stepped out of a Damon Runyan novel. When he spoke, you didn't have to guess where he was from – New York City. Tommy was an ex-Marine who, like many others, carry their training and loyalty like a decoration. He was less than six feet tall, solidly built, and very tough. He always wore a Marine Corps belt buckle and cap with the word "Marine" on it. Tommy was both profane and funny. He told a story a minute, most of which I didn't believe until the rest of the crew informed me that for the most part the stories were true. Most people are not very good at cursing; not so with Tommy. It didn't offend me, for it seemed a natural part of his makeup. He could put a skilled Arabian camel driver to shame. His peers made no secret of the fact that they believe Tommy is a top notch deckhand and that few could match his skills.

By six p.m., we were at the Metro Oil Terminal in Rensselaer, New York.

It took us 12 hours to pump out our water ballast and refill the barge with oil. After supper, Marty Kehoe (owner of the tug) came aboard to pick up the radio which needed repair, and stayed until midnight. Marty would never admit it, but I think he gets lonely and needs to talk with the men. In any event, the sea stories came thick and fast in the galley. I'm not sure if they were trying to impress me or Marty, or maybe both of us. I enjoyed every minute of it. A large part of these trips is sharing in the camaraderie of the crew.

Typical of the stories told was this one. A deckhand had a hawser to splice on a cold and rainy day and decided to bring it in the galley to work on. The cook was not happy with the idea, but was tolerant because of the weather. The deckhand brought one end in, then went out for the other end and his splicing tools. Squatting on the deck in the warm and dry galley, he quickly went at his task. From time to time, the cook would give him a strange look, but he figured the cook was still annoyed at him for cluttering up his galley. Polishing off the splice and a cup of coffee, he started to drag the hawser outside. In a flash came a rude awakening, punctuated by a belly laugh from the cook. My God! He'd never live this down. He had brought each end of the hawser through opposite sides of the galley, resulting in a perfect loop through the galley and around the end of the deckhouse. Leaping to the bulkhead, he grabbed a fire axe and a block of wood. With one savage stroke, he severed the hawser, dragged it outside, and ignoring the rain, re-spliced the blankety-blank rope in record time.

I was up at five a.m. the next morning, a Sunday, ate a huge breakfast and then sat in the wheelhouse with Captain Amos Yell and Dave Oliver, the Chief Engineer, talking as we waited for the fog to lift. It was fall, and fog is a constant problem at this time of the year. By 10 a.m., the fog has lifted enough for us to get under way. Amos was on watch and Larry Pauquette, a deckhand, was sanding down the capstan. A capstan is a powered winch that is used to pay out or haul in the towing hawser. The hawser is not wound directly on the spool, but on blocks of hard wood that are attached to the spool. They wear out but can be replaced quickly. Larry was rubbing linseed oil in when I asked why that instead of paint to protect the wood. He patiently explained that if you painted the blocks, the hawser would stick and there are times when you want the hawser to slip smoothly rather than grab.

Monday started out to be a beautiful sunny day and stayed that way, giving me an opportunity to shoot lots of pictures of the crew. Although the pictures are primarily for my book, I always send copies to each crew member. I also spent time on the barge taking pictures. The barge we pushed this trip was the FULTON. She was liked by the bargemen because she was new and has excellent, air-conditioned quarters. The tug crew aren't too fond of her because she was built cheaply with thin sides and is very easy to damage.

We entered Lock 2 at 3 p.m. and one of the deckhands had to climb the wall to place lines. At 18½ feet, it's just too high to heave a line. The same thing happened at the next lock, Lock 3, which at 19 feet is the highest lock on the Champlain Branch. Jimmy Day was on watch. It was very peaceful as we listened to soft music on his radio. This is not the norm for Jimmy. Usually the radio would be tuned to country-western and we would be listening to Jimmy singing along.

After supper was cleared away, some of the off watch and our cook, Geno Pallozzi, were in the galley relaxing in the same way you might at home. They were reading the Sunday papers, half watching T.V. and hassling each other. I wandered up to the wheelhouse to talk with Amos. He was sweating out an early fog. It might hit us as early as Lock 5. Amos managed to get to the Guard Lock above Lock 6 before it became too thick to move safely.

It was still foggy the next morning so Bob let Geno sleep in and each of us got our own breakfast. During breakfast, Tommy Davis talked about how cheap our other engineer is. He said he was so cheap that if a trip around the world cost a dime, he wouldn't get out of sight of his house. This led Tommy into a true story about Larry, our other deckhand. Larry was in a gin mill where the music from the juke box was a little too loud to suit him. Larry asked the bartender to turn it down, but he got a negative answer. Larry slowly walked over to the offending machine and proceeded to pour his beer down the coin slot. With sparks flying, smoke billowing out and confusion reigning, Larry took his leave. Tommy forgot to tell the rest of the story. Larry said Tommy and two others were in the same bar later that night, got into an argument with some native sons, and almost demolished the place. At times boatmen do play a little rough.

We reached Lock 7 about eight p.m. and a crew change took place. We dropped one bargeman off and picked up barge captain John Dalton from Nassau, New York. At each lock I like to watch the crews handle lines, and this lock was no exception. It always surprises me when they miss what I would call an easy throw. I compare them with a rodeo cowboy and they shouldn't ever miss. Maybe it's because I have seen them make impossible throws. As soon as the lines were in, I walked to the galley for coffee and my timing was perfect. Geno had just taken some blueberry muffins from the oven. I admitted to having a radar nose and savored the offered muffin.

I observed another unusual navigational aid that morning. Amos was making a security call – "ERIN KEHOE passing Potato House Northbound." It really is a potato house on a farm. Many of the crews have an intimate knowledge about the houses and farms that line the edge of the canal. An example is the horse farm near the potato house. Amos automatically slows down because he knows it's shallow and, sure enough, we start churning up mud and are almost on bottom. This is comparable to the pilots on the Mississippi River who also have to know every inch of the river. It's a hazard deep water sailors don't have to contend with.

As we got closer to Whitehall, Amos was on the CB radio to his wife. Using the CB is a way of getting messages to families of the crews. Many times there are requests for things to be brought to the boats. The wives meet the boats all over the canal. We approached Lock 9 around lunch time, and we were stopped by our first red light on this trip. We have to tie up to the wall and wait. While waiting, I note the sun is still out, but a cold wind is coming out of the West. The thought crosses my mind that it will be a cold night on the lake. I wouldn't have given the weather that much thought had I been home. I'm beginning to think like a boatman. Twenty minutes later, the gates slowly creep open to reveal a large trawler yacht, probably headed for Florida for the winter. A friendly wave in passing as we get the green light and start into the lock. We are at the "top of the hill" – highest point, and from now on we will be locking down. Dave Oliver, our Chief Engineer, got off here for a few hours. We will pick him up at Lock 11 in about three hours. Larry takes the tow out of the lock, giving Amos a chance to eat as our mate isn't back aboard yet to relieve him.

Our two deckhands, Larry and Tom, constantly hassle each other using language that would peel paint. It's all good natured and you can tell they like and respect one another. I've noticed that the crews rarely talk about worldly topics. They do gobble up newspapers, but their reading usually runs to sports and news related to their job. Women don't have a corner on gossip. Hardly a day goes by that they don't bend my ear about someone. I also hear gossip from their boss, Marty Kehoe. I felt good being able to share in their candid conversation, for it meant they trusted me. I was accepted as one of them.

Dave Oliver and Jimmy Day are waiting for us as we make Lock 11. Jimmy is very proud of the binoculars he has just bought as they are a symbol of his job as mate. Between Locks 11 and 12, we received an "air mail" delivery. We needed Freon for the refrigerator and had no place to tie up. The Chief Engineer didn't want to wait until we reached Lock 12, so Paul Gordon stood on a bridge as Larry climbed to the top of the wheelhouse. Timing it perfectly, Paul dropped the package and Larry made a Willy Mays's catch amid much cheering from the crew.

Jimmy has the wheel now. He brought his tape recorder aboard and was playing his country-western music favorites. Jimmy had been to the hospital to see his mother who was much improved. This on top of Jimmy's normally good spirits put him in low orbit. His good spirits were infectious. Geno was in the wheelhouse, too. He is one of the few cooks I've seen that spend any time here. It's not that it's off limits – they just don't do it very often. Milt, an engineer, was filling the aft ballast tank before we got to the lake. This will reduce the vibration on the stern, and also give us more power.

The laundry was waiting for us at Whitehall. So were the families of Amos, Paul and Dave. A few minutes were spent together before we had to start out again. While eating supper, I watched Geno washing dishes. The sinks are so deep he has to bend double to do the dishes. It seemed to me as if one of them should be shallow. That's what comes of being an Industrial Engineer: I keep looking for ways to improve an operation. In this same vein, I wanted to see how the engineers coped with the noise of the engine room, for long exposure can lead to hearing impairment. Some engineers used store-bought attenuators, some used cotton, but I think, for availability and price, Dave Oliver chose the best – toilet paper!

We arrived at Plattsburg in time for the watch to change, but Jimmy took it in so Amos could get some much-needed sleep. This evened up for Amos steering for Jimmy the day before while he visited his mother in the hospital. Doing favors for one another is common on the ERIN KEHOE. It was a cold morning so Larry and Tommy were having a good-natured argument as to who should tie up the barge as there is only one warm jacket between them. Since it's Tommy's jacket, well . . . Larry lets Tommy get started, then comes out to help put away the searchlight cables and take in the safety lines.

There's an interesting feature of travel on Lake Champlain. The lake starts to widen out at Split Rock Point. At this point, if the weather looks threatening, the cables that tie the barge and tug together are taken off, and they are run with the safety lines only. In rough weather, the cables would part because of rocking between the tug and barge. Once you start around Split Rock, you're committed: it would be darn near impossible to turn back.

While the barge was being pumped out, everyone was asleep except Tommy, Larry, and me. We were in the galley being regaled by Tommy with another of his escapades. It was about Tommy getting knocked out and then the guy did a number on his ribs." Larry dutifully asked if any of his ribs got broken. Tommy said, "Hell, I've been kicked so many times, my ribs bend instead of break."

After his story, the normal work routine started with Larry washing the boat down from top to bottom and Dave painting the engine room. With the breakfast dishes put away, Geno and I walked up to a small store to buy something for dessert because the oven is tied up with a turkey. Being without dessert is a rare occurrence for Geno. He is one cook who prides himself on his homemade desserts.

It was a cool but bright beautiful day, and after lunch, both deckhands talking about who in the company will be bumped (laid off) if Marty Kehoe decides not to run the JAMES when the ice starts to form. The hull on the JAMES is paper thin and it would take very little to punch a hole in her.

Incidentally, I should point out that we did not dock in Plattsburg proper, but a few miles south of the town. There was a good reason for this: Bob Gordon doesn't like to tie up too near town because of the possibility of a drinking problem cropping up with some of the crew.

Later in the day, I spent some time with the young lad, Victor Stockwell, from the barge. This was his first year, and he spoke about how difficult it was to get some barge captains to teach him the trade. His connection to the business is his father-in-law who is on an ocean-going barge. Family is the most common way of breaking into the business. He was lucky on this trip because the barge captain was John Dalton, an excellent and willing teacher. I also liked John for his feelings about Marty Kehoe. John thinks highly of Marty because he is always bragging up his men. This builds up loyalty among Kehoe men.

Even though we were not quite finished pumping out, at four p.m., Bob decides to start making up the tow. In less than an hour, the safety lines are on and we were heading down the lake. Waiting for supper as the sun went down, we found a cool breeze giving the lake a slight chop. Standing out of the breeze I felt the sun warm my face, yet I could taste the freshness in the air. You can build a natural high under these conditions. I envy the crew and wish I had stayed in the business. My notes showed that we had a super supper: pork chops, yams, salad, green beans, home fries, applesauce, and custard pie.

That night, Jimmy had a rough time because of the fog, and he finally was forced to shove the barge against a mud bank as there was no place to tie up. It was three a.m. and just three miles from Whitehall. Jimmy started out again at 7:30 and on the bend just beyond South Bay, we met and passed the SALUTATION headed North. This gives me the opportunity to explain the unique method tows use to pass each other when the water is both shallow and narrow. They head directly toward each other until you think they will hit head on. At the last possible moment, they turn out and slide past. If they get over too soon, the suction and currents will put the barges into each other.

It was Wednesday and therefore crew change day. The only member to get on at Whitehall was my friend and Chief Engineer, Paul Gordon. He wanted me to take note that he had relieved Dave Oliver early. Paul, of course, was joking, but technically noon is the official time to change, which very rarely happens. The rest of the crew got on at the next lock. In less than five minutes, I again felt a part of the oncoming crew, even though I had never met any of them. They were a study in contrasts.

The first two were deckhands Diz McCloskey, who at 52 looked 65, with a boatman's belly, and engineer Garry Kase, age 31 looking 21, short and slight built. The next two were cousins Don Chuput, deckhand, and Bob Nolin, the cook. The last man aboard was the first boatman to look like everyone imagines a sailor should look. He was dressed in a white cable knit sweater, bare head, of course, and ruggedly handsome with a thick shock of slightly graying hair. He swung aboard as if he owned the world, with a look that dared anyone to ask to see his title to it. I was impressed and a little in awe of George Murphy, our mate. I could sense he was not keen on having me for a roommate. Being a captain, although not on this tug, he was not used to sharing his quarters. Shortly I was to wonder why a captain for the world's largest towing company (Moran) would be working as a mate for a small outfit such as Kehoe. Well, it was none of my business, so I didn't ask any questions. I'm glad I didn't ask, because later I saw a sample of George's fine Irish temper. Being of the same persuasion, I am a good judge of it. Unfortunately, it was aimed at me, not once but twice. And I deserved it both times. I can't believe how stupid I acted. More about that later.

While passing under the Dewey bridge near Comstock, the crew told

Chief Engineer Paul Gordon and Deckhand Don Chaput—"Making Repairs."

Rigging up to unload oil on Lake Champlain, Bill Tungate (fore), Paul Clamp (aft).

me this was the infamous bridge that has reached out many times and decapitated wheelhouses. It got three Kehoe tugs.

I woke up Thursday morning to find us tied to the Metro Oil dock at Rennsselaer. No one was up but me. It was dark and deathly quiet as I fixed myself breakfast, with only the rumble of the generator engine to keep me company. By 6:30, some of the crew had started to stir. It was a slow day as we waited for the barge to be loaded.

I did get a chance to talk to George Murphy, and I got a surprise answer to my first question. As a boy, he lived a few blocks from me in Lansingburgh (North Troy). We never knew each other because he was a couple of years younger. Even a couple of years can make a huge gap as far as who your friends are when you are young. His career didn't have a very auspicious beginning. While he was on the MARY O'RIORDEN heading for New York from Albany, she broke down four times. They couldn't make much headway against an incoming tide, so his first trip took many times the normal time to get to New York. Like many others, his connection to the business came through an older brother. George learned the business in some of the old outfits that are long gone. He spent time hauling molasses for Connors. More time with Russell, Blue Line and then he earned his mate and master's ticket while he was with Bushey. One unique assignment was as Captain of the tug BOSTON, the first automated tug in the New York area. Just before coming with Kehoe, both George and his brother were Connecticut River pilots for Moran.

This trip we were plagued by fog every night. This gave George a great break because the fog always hit us on his watch, which meant George got lots of rest. Of course, this led to some ribbing by our Irish deckhand, Diz McCloskey, at meal times. Diz, with Gaelic good humor, would suggest that George come around once in a while and reintroduce himself to the crew. Diz also showed great concern that George might be developing a back problem from spending so much time in the sack. I think George tolerated the ribbing from Diz because of Diz's position in the crew and the fact that Diz was Irish. I don't think any of the rest of the crew would have taken a chance on touching off George's Irish temper.

His temper was awesome, and I saw it twice when it was aimed directly at me. The first time was when I got in his line of sight as he was lining up the tow to enter the lock. I had done the same thing to Amos Yell on my first trip ever. But the second incident with George was far more serious and frightening than just getting in his line of sight. I committed a cardinal sin for a canal boater – I went into our room after George had gone to

"Approaching the lock."

"Close to nature."

61

bed. To compound the sin, as I opened the door, the sun struck him smack in the eyes! My God, you never saw such a reaction. I had bent down to get something from the drawer under the bunk, and my eyes were three inches from his when they flew open. He only yelled five words, but the threat behind them was real and chilling. "I'll get off the boat!" he said. Now it is important to understand that this seemingly bland expression carried a tremendous invective, for in his customary simple canaler's argot he really was saying, "If you don't get the d-- clown O'Malley off this boat, I'm leaving!" He was already halfway off the bunk before I tore the door open and escaped.

I was positive George would get violent and throw me over the side. I ran to the stern to think, pondering my fate. At the very least, I felt, he would insist that Captain Bob Gordon put me off at the next lock. I had two alternatives: I could slit my throat right then and there, or I could figure out some way to avoid George for the rest of the trip. That, my friends, is no easy task when you're trapped on a vessel only 81 feet long.

I avoided the now-riled Irishman until he sat down to lunch, and then I quickly pulled my gear from our room and bedded down with the crew in the forepeak, where I should have been in the first place. What made the incident so stupid on my part was the fact that I knew the rules and forgot them. When two men share a room, one always removes everything he might need before going on watch, and he *never* re-enters the room during the other's sleep time. In foul weather, you can pass through a tug without going on deck. The rooms are connected for this purpose – but they are in line, with men on the same watch assigned rooms on the starboard side and likewise on the port side. The only time there can be a problem is when a boat carries a passenger, who obviously doesn't stand a watch.

I prefer the forepeak because there is more room. It's warmer in cold weather and cooler in hot weather. Of course, I apologized to George and he accepted it. We Irish are quick to blow our cool, but just as quick to forgive and forget. I carefully avoided making any more dumb mistakes on this trip.

The rest of the trip North was uneventful except for two seagulls looking for handouts that escorted us all the way to Plattsburg like a couple of winged sentries.

At night when I go to bed, I always have a companion next to me. Dark and brooding with a pair of violet eyes that glow eerily in the dark. It's the gyro for the compass and it sits next to my bunk. It looks like some-

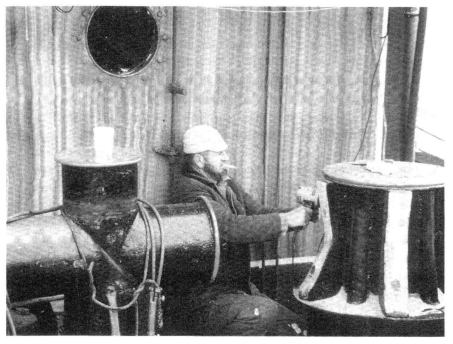

Larry Paquette sanding the wood blocks on the capstan.

Bargemen brothers Bill and Harry Tungate take a break.

thing out of a science fiction movie. It sits there humming, brooding and contemplating who knows what. So far it hasn't attacked me!

My log is blank for a day and a half. The next entry shows us southbound and arriving at Lock 7 where we were delayed by the MOBIL CHAMPLAIN, which was running on one engine. I should explain. The MOBIL CHAMPLAIN is a motorship, that is, she's like a motorized barge. Instead of one engine in the hull, she has two outboard engines such as you might see on a pleasure boat, except they are much larger. The crew says they have a lot of problems with these engines. I noticed they were broke down in Waterford when we were coming up. We also had to wait for a sister ship, the JAMES KEHOE. I was very glad to see her because a friend, Harry Tungate, was on the barge she was pushing. I only had time to wave and yell a greeting before they were past us. Even though I had taken several trips on this tug, I seldom see some of the crew more than once. The reason is the shifting of crews by Marty Kehoe and/or the men may be on their time off when I take a trip. About the only one I can be sure of is Captain Bob Gordon.

As this trip wound down, I noticed that each afternoon, deckhand Diz McCloskey brings mate George Murphy a cup of coffee and takes a break himself. Although he doesn't need it, I observed that Diz usually had a piece of pie or any other sweet he could find. As a nice gesture, he invites the young lad, Victor Stockwell, from the barge to share his coffee break.

I took a short nap one afternoon and as usual, after getting up, I headed for the galley. Bob Nolin was giving his cousin, Don Chaput, a hard time about his going into the refrigerator. Bob says he's not running a snack shop. This leads to a discussion that presents a dilemma to Bob or any other cook for that matter. The grub belongs to the crew, yet the cook is responsible in making sure they don't run out of food.

After George Murphy started talking to me again, he gave me a Murphy rule for prudent dock operations. He said you never whistle for an oil dockman as you approach the dock. First you land your barge, tie it up, and then whistle. That way if there is any damage to the dock, you can claim someone else did it.

I scratched one last note as we neared Lock 1, where I would leave the ERIN once again. It was about Bob Nolin's start as a cook. Although he has worked for many outfits such as Sears, Morania, and Great Lakes Dredge, his start was a little shaky. He was hired by New York State to cook on a five-man state tug. With no experience and a cookbook under his arm, he reported to work only to discover he was assigned to a 26-man

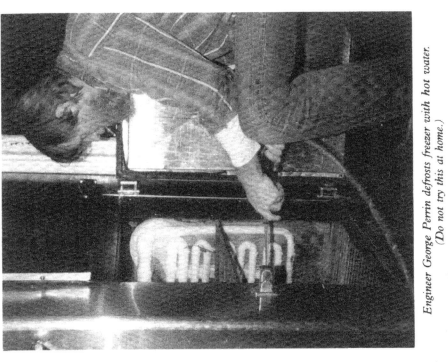

*Engineer George Perrin defrosts freezer with hot water.
(Do not try this at home.)*

Otto Andereasen oversees greenhorn Jerry Golden in art of tying knots.

dredge and not a five-man tug. He survived, but barely, to become a good cook. Like most cooks, he hates to shop, but likes to see the men enjoy his cooking. I once asked him if he ever served hash. His answer was an emphatic "No!" One reason is the boat has no meat grinder, but even if it did, he would serve hash only as a side dish, if at all. But Nolin does make truly great coleslaw.

With this note, I close my log on another trip on the ERIN KEHOE. All in all, it was an interesting experience sharing a few days with Amos. Yell, Bob Gordon, and the other boaters – even George Murphy, who probably was still wondering why he ever signed on as mate with Marty Kehoe and drew me to share a room with.

Larry Pauquette, Tommy Davis and Dave Oliver take up the slack.

The ERIN KEHOE on Lake Champlain.

67

Captain Bob Gordon catches up on his paper work.

Cook Geno Pallozzi setting the table for supper.

CHAPTER 5

ON THE ERIN KEHOE – 1979

I began my next trip on the ERIN KEHOE by spending a lovely afternoon with my old friends in Waterford, Captain and Mrs. Jack Maloney. Not knowing exactly where the ERIN might be, I left the Maloneys at 4:00 p.m. and headed for Lock 2. Not finding the boat there, I kept on until I found it going into Lock 4 at Stillwater. I talked a few minutes with Mate George Murphy and asked him when he thought they would be in Lock 1 at Waterford. His estimate was 6:00 p.m., and he only missed it by twenty minutes.

I drove back to Lock 1 to wait for Captain Bob Gordon. A few minutes later he arrived, and we decided we had enough time to run down the road for a cup of coffee, before the ERIN KEHOE would be in Lock 1. While doing this, he could bring me up to date on what was happening that year with all the men I had traveled with on other trips on the ERIN.

It was cold and raining as Bob and I stepped aboard on this 9th day of October, 1979. I was thinking that one of these times I was going to get smart and take a trip when the weather was warm and sunny.

Each trip begins with a little suspense for me. There are always some crew members I have never met. This time it was a first-year deckhand, Joey Aurarello, from an old Waterford, New York boating family. The cook would have two extra mouths to feed this trip. Besides me, the other one was a young lad, Jerry Golden, from New York City, who had been riding (he received no pay) for four weeks to pick up some deckhand skills. It was a rather odd set up until Bob explained that Jerry's dad and Marty Kehoe were friends, and Jerry's dad thought some time on the tug might help the lad "find himself," as young people say today. I don't think Jerry appreciated the rare opportunity he was being offered. It was a very difficult business to break into, if you do not have family already in it. Unfortunately, the consensus of the crew was that he wouldn't make it. I remember hoping they'd be wrong. One thing he did do to perfection, however, was eating. Wow! Could he pack it away – almost as well as I could at his age.

The last man I met that day was the cook, a new man who was on his first week of cooking. Although he was not experienced, the crew was tolerant while he learned. Their patience was stretched to the limit one morning, however, when he made the pancakes in a greasy frying pan.

Tommy Davis. Painting goes on forever.

Tourists at Lock 9.

In place of a mate that would normally be carried aboard, Captain Amos Yell was moonlighting from the Bouchard Transportation Company. Amos was a friend from my first trip on the ERIN. I was amused to see Amos get mad at himself. It happened when he had a little trouble putting the tow into the Federal Lock at Troy, New York. He adjusted well for the current, but the wind caught the empty barge. It was a rare sight to see Amos Yell getting hung up at a lock.

Later, I was all alone in the wheelhouse, with Captain Bob Gordon steering from the dog house (upper wheelhouse) because we were running with a light (empty) barge. He wouldn't be able to see over the barge if he were standing where I was. As soon as we left Troy and had no bridge clearance problems, deckhand, Otto Andereasen put our mast up. Since we were now in Federal waters, we had to carry lights to show other vessels that we are a tug pushing a barge. This is indicated by two white lights, one above the other on the steel mast. Otto's next job was to rig the clothes line to each searchlight so they could be controlled from the wheelhouse.

Next morning by 7:45 a.m., there was a misty rain and fog, and visibility was getting worse by the minute. With a busted radar, we knew we would have to tie up somewhere soon. By 8:30 a.m., it became too dangerous to move, so we tied up to an oil dock (owner unknown) at Verplanks Point on the Hudson River. The fog was now so thick we couldn't see across the river to Stony Point. Nine o'clock and it's starting to snow! For the love of Pete, we thought, it's barely into October. At noon, we're still tied up. Lunch was corned beef, boiled potatoes, carrots and jello – not exactly a winning meal! Some of the crew here getting restless, for this was change day for part of the crew, and we were supposed to be in Port Reading, New Jersey, by 4:00 p.m. Instead, there we sat at two o'clock in a blinding snow storm.

By 4:30 it cleared enough so that Amos decided to try it. Both barge and tug were almost out of grub. The outside temperature was 39°, and we were running in a miserable driving rain. The cook was busy with trying to scrape up enough food for supper. All in all, it was a boring day for everyone. The crew slept a lot, especially our chief engineer, Dave Oliver, who was alone in the engine room. The other engineer, Jack Vetter, was waiting in New Jersey for us. Everyone aboard was wondering if Geno Pallozzi, the regular cook, who also was in New Jersey waiting for us, would be in a foul mood. He's never happy about waiting or traveling to New Jersey to pick up the boat.

Supper wasn't too bad considering the state of our larder. We had

Larry Paquette, "walking the dog."

Victor Stockwell, "walking the dog."

72

hamburg, homemade soup (a little bit of everything the cook could find), potatoes, salad, and ice cream.

Captain Bob couldn't get through to Ecklof's yard on the radio to get a message to Geno to buy grub, and there was no where to get ashore to use a phone. While we were tied up, some of the employees of the oil company drove to the top of the hill to see who we were. There was always the possibility they would contact Kehoe and claim we damaged their dock.

The deckhands put on what I call a canvas band-aid. They rigged a strip of canvas around the wheelhouse and halfway up on the windows. This makeshift arrangement served two purposes: it kept out drafts and stopped the rain from dripping on one of the bunks below. Old boats are like old houses they need constant attention.

It was lousy timing. The fog smothered us just as we were tying up, and it took all of Bob's skill not to smash the tug into the dock. The cold and dampness went beyond the bones that night, but the galley was a cozy haven to sit with friends, drink endless cups of coffee, and tell lies.

Breakfast Thursday morning was a thousand times better with Geno Palozzi in command of the galley. Shortly after breakfast, I met Jack Vetter, the ERIN's other engineer. Two things I remember about Jack, The first was that he lived in Florida and commuted twice a month to this job. That has to be some kind of record for distance traveled for a job. secondly, there was the slight smile and twinkle in this eyes that made him appear as if he held some great secret or was about to play a practical joke on someone. Whenever Jack visited the wheelhouse, he always wore his noise protectors high on his head. This gave him the look of a mouseketeer from the old Mickey Mouse television show.

Bob Gordon called Marty Kehoe to see if he could get the radar fixed. Bob got a firm no. Oh well, such turndowns were all in a normal day's routine in the tug business. Then Geno started complaining of sitting in Ecklof's yard all day. Geno said the Captain could have used the radio to let him know how late we would be. I guess Geno didn't believe we tried to reach the yard on the radio.

Gradually, at last, things got back to normal. Bob Gordon had his usual bowl of soup at 10:00 a.m. Young Joe had the wheel as we passed midtown Manhattan on a near empty river. Lunch was terrific: bean soup, excellent steak, and banana cream pie. You might be thinking how the devil we could have all those goodies when we were nearly out of grub. The answer is simple. Geno, being the veteran he is, knew we had to be low and like all good cooks, he carried extra money for just such an emer-

Tower between Albany and Troy, New York greets us with a wink.

Engineer Bill Spring in his normal steaming position.

74

gency. He also was considerate with little touches like putting ice in everyone's glass so we didn't fall all over one another going to the freezer.

At ten past eight on an overcast day with the temperature in the mid-40's, I was sitting in the galley of the ERIN KEHOE having a cup of coffee and observing Geno. He was very quiet as he mixed up pudding. He had had a heart attack since I last saw him, and it left him changed. It seemed as if the spark had gone out and it left him very moody. It may be normal behavior after what he has been through, but I kept hoping he'd snap out of it. I thought he was scared, but I didn't know what to say to help him. It's a terribly helpless feeling.

It was a perfect afternoon to take a nap, and that's just what I did. When I got up, I noticed the barge was drawing power from the tug. When the generators are running on this barge, the exhaust blows in the wheelhouse windows, and diesel fumes are not exactly Chanel #5. A good reason tug crews don't like this barge.

We expected to make the Federal Lock at Troy by 7:30 the next morning. The boat was vibrating so badly that night I could hardly write my notes. Our ETA was not too far off, and we were at the approach wall at 7:00 a.m. After that, it took only 20 minutes to clear the lock. Traffic was so light we rarely had to wait at this or any others of the canal locks.

We eased into Lock 1 of the Champlain Branch, an hour and a half later. It was time to get rid of our garbage and for Bob to call Marty Kehoe and the barge captain to call Ecklof to report our position. I noticed that the boatmen never seemed to rush to get through the canal locks like they did in the Federal Lock at Troy. They moved at a comfortable pace and took a few minutes to talk to the lock operators, many of whom they had known for years. I think it's good that they aren't completely dominated by our modern mania called pressure, to move it, move it, move it.

Mid-morning and it was still raining. Looking at the notes I jotted down the day before, I wrote,that on the river and lake the bargemen and the tug deckhands get a break from work (especially in foul weather) if they are caught up with the endless painting and polishing. Of course this is not true for the men in the wheelhouse and engine room. And like the proverbial housewife, the cook's work never stops.

It kept raining all afternoon. Cold weather may sap one's strength but worse to me is wet weather that saps one's spirit. This seemed to be the case in the wheelhouse that day with Amos, Otto, and me this dank dismal afternoon. Conversation was sparse and certainly nothing earth shattering was brought up. Otto, our dour deckhand, rarely contributed even when

Tug MORANIA #6, going into the notch, Lyons, New York.

Mate Jimmy Day.

76

the weather was fine, but he listened with the same intensity that he performed his work.

There is a bright spot in all this; first it shows that tugboating at times can be boring and tedious as any other job. There are times when you have to grit your teeth and make it though the rest of the watch.

Because conversation was slow, I observed Amos more closely than usual. Each man handles the wheel differently. While steering, Amos, in contrast to Bob Gordon, kept the wheel in constant motion. It wasn't a conscious effort on his part, but almost a reflex action, and the wheel moves just a degree or two in each direction. Just that whisker of movement kept the tow dead in the middle of the channel.

We hit Lock 5 at 3:00 p.m. and the barge men had to climb the lock wall to throw a line down to Otto. The wall was too high for a deckhand to heave a line to the top. It was pitch dark as we approached Lock 8 at 8:30. Captain Bob turned the searchlight just long enough to get the tow started into the lock. Then he had to turn the light off quickly before be blinded the line handlers.

The MARTIN KEHOE called us on the radio to ask if we could raise the operator at Lock 12. They couldn't reach him and they needed a refrigerator mechanic to meet them at the lock. Amos couldn't reach Lock 12 either, so he called his wife on his CB radio, and she called the lock operator on the telephone. The fog shut us down just short of Lock 12 a little after one o'clock. Sometimes when boats are tied up for fog and it's unlikely they will get an early start, the captain elects to let the cook sleep in. Such was the case here, so Geno got some much needed extra rest, and we all made our own breakfast.

Miserably cold it was when we got up, but by ten o'clock it was comfortable enough for Otto to take the young lads to the fantail to teach them knots and line handling. Otto's a good teacher, and he showed great patience. Since Marty didn't seem interested in getting the radar fixed, Dave Oliver, the chef engineer, had a go at it. Meanwhile, Geno fixed another excellent lunch of pork chops, spinach, yellow beans, applesauce, potatoes, and cake.

One of the barge men, Jerry Hanly, told me a fine story about one of Bushey's men. This man had a habit of leaving the tug in a rather unorthodox way. In many places, there is no way to get the tow close enough to shore for anyone to step off. Bushey's man solved this by standing on top of the wheelhouse and having the tug slow down just before reaching a bridge. At this time he would grab a girder and swing up on the bridge.

Daily ritual, 10 o'clock soup for Captain Bob Gordon. George Perrin joins him with a problem.

Captain Amos Yell, Jr.

Once, though, this action came out a tiny bit different. This time he did his thing and a couple of minutes later, a deckhand looked back and damn near went into shock. There were a pair of legs dangling under the bridge! Because of a miscalculation in the tug's speed, our man was pasted against the girder. Luckily his death grip on the steel saved him until he got his breath back. My friend Rod Stafford did a fine job drawing an illustration for this story.

Between 9:00 and 10:00 we met the tugs MARTIN KEHOE and CHEMUNG coming down. These were the first tows we had seen on this trip, which seemed to have little traffic. We made Port Douglas at 6:30 p.m., and it was too late to call Mrs. Murphy's taxi service to take Geno to town to shop. Mrs. Murphy is a phenomenon, the likes of whom are found in towns too small to support a bus line. She's a feisty little woman who always wears a chauffeur's cap and knows everybody and everything that happens in her town. Once when Charlie Passero and I went for grub, he insisted I sit in front with her. He was a little in awe of her and apparently had done something to upset her. I enjoyed her conversation and the cooks always appreciated the fact that she was accommodating and reasonable with her service.

This particular evening was a very quiet one. No one went to town, as we expected to leave very early in the morning. And early it was - 4:00 a.m.! There was some concern that the supermarkets wouldn't be open in Whitehall on Sunday, for we really needed grub. By 7:00, it seemed that it was going to be another lousy day, weatherwise. I had hoped for a bonus on this trip, and that was to find some good color on the trees. But it was not to be. Fall foliage in this part of the country can be spectacular. I call it calendar country land everyone should see it.

We had tried to raise the MARTIN KEHOE on the radio the day before with no luck, and we found why when we got to Lock 12. While tied up in the fog, someone came aboard and stole their radio and the antenna. It was lucky for the thief that none of the crew woke up, and Marty Kehoe was not going to be too happy to hear of the theft from his boat.

Winter certainly was on the way, as just before lunch, we saw a large flock of Canadian geese flying south. Then by mid-afternoon, we were in Whitehall and found the Grand Union was open. Ten bags of food that day came to $170. Bob's wife brought Geno and the grub to Lock 11 for us – another example of how boating families help out.

At five o'clock, we were held up at Lock 8 waiting for two tows to lock up. One was a Bushey boat (CREE 2) and the other was a tug owned by

79

Deckhand Dick Meier makes a perfect throw.

Captain George Murphy.

the man who owns the barge we were pushing. We had another wait at Lock 7 for a different reason. The state has cut back on its budget for operating the canal system so that lock and bridge operators have to take care of more than one structure [a bridge or lock].

I seemed to pay more attention to observing mannerisms on the tug than I do at home. For example, when anyone talked with Dave Oliver, Dave always shifted his Navy watch cap from the top of this head to a spot almost covering his eyes. When the conversation was over, the cap went back to the top of his head. Dave was also very camera shy, and there are consequently few photos of him.

Some evenings when we were running with a light barge and the control was in the upper wheelhouse, I liked to sit in the lower one in the dark to relax. There I could sit, dream, fantasize that it is me and not Bob Gordon or Amos Yell who captained this tug. I could feel the response of the wheel and the surge of power as I take over the watch. I used to get the same feeling when I would test run aircraft engines on fighters and bombers during World War II. I also could imagine the envy generated on shore by all those unseen people because they couldn't be a part of this boater's world. It's a euphoria I could never experience on my regular job. Boatmen will generally deny it, but there is a romantic facet attached to their line of work. The only sounds to accompany my racing imagination at such times is the soft whirr of the gears as the wheel turns. Even the few lights that shatter the darkness help to feed my fantasies. They are the tiny lights on the rudder position indicator, radio, binnacle along with the four red bulbs on the power panel. The muffled hum and vibration of the engine seems to set up a rhythm that says I belong here.

Sixty feet aft from where I sat the young lads finally gave up on the TV. It was too full of static so now they were arguing the merits of pop songs and CB radios. My escape to the solitude of the wheelhouse was all the more rewarding under the circumstances.

Monday morning we were once again caught by fog. At Lock 2 we lost four hours before the fog cleared enough to make it safe to continue. We cleared the Federal Lock at seven o'clock on another cold, damp, and cloudy morning. Nonetheless, courtesy prevailed with this crew. We slowed down to pass the Albany Yacht Basin so our wake wouldn't damage the boats. I stress the word *courtesy* because all boats do not follow the rules, and that gives tugboat crews a bad reputation. Bob refueled the ERIN at the Albany Mobil dock and while there, MORANIA 12 also approached for refueling. This gave me an opportunity to take a photograph of their deck-

"Contrasts, Manhattan from the Hudson River."

hand waiting to heave a line – he was a friendly, bearded giant of an Irishman, named Jim Kelly. As we left the Mobil dock, it began to pour, so I headed for my favorite place in this kind of weather, the galley.

As the south wind picked up, it built some choppy water and the tug really began to bounce around, but by four o'clock the wind had dropped and it became a little warmer. More ballast was being pumped out of the barge, which had risen so high that Amos had to place his stool on top of a pallet in order to see enough to steer.

Tuesday morning we were at Ecklof's yard in Staten Island. We were switching barges and taking out the E-20. We had left the E-19 in a slip at Shooters Island. In the confusion Monday night we left our electric cables on her. We had to retrieve them before we headed north again. The confusion (and I might add the frustration) began for Amos when Ecklof's dispatcher didn't know which slip we were to drop the barge into. The barge men didn't know either, so Amos took an educated guess that turned out to be the right one. The second act began as the port search light went out as Amos was ready to land the barge. The finale came when the barge men wanted to put out a line where Amos didn't want it. This was a bit upsetting for the captain. The air turned a brilliant blue as Amos spoke to our deckhand, Otto, and told him to put the damned line where he wanted it and not where the stupid blankety-blank barge men wanted it.

There was a bonus for me when we switched to the E-20, for I found that a friend from earlier trips, Bill Tungate, was aboard. Both Bill and his brother Harry are good boatmen and great companions.

Then, the sun came out for the first time since I had come aboard. It's difficult to convey how I felt at the time, but it is exciting moving through the Kills on a warm beautiful morning with dozens of tugs and ships in motion. About nine a.m., we passed the Statue of Liberty and, as usual, I had to take pictures of her. No matter how many I take, I'm never satisfied. We moved past the Battery and the traffic began to disappear. Between the Battery and the Tappan Zee bridge, we met only one tow, the ROCK-LAND COUNTY, with its usual contingent of trap rock barges.

We ran close up to the west shore of the Hudson River, and this gave me a chance to see up close a floating restaurant I was curious about. It used to be the Binghamton, an old Erie-Lackawanna Railroad ferry. It is tied up just south of the George Washington Bridge.

Although I tried to take a nap in the afternoon, I had to give up the idea. When conditions are right, vibrations on a tug are potent, and this was one of those times. The rest of the trip was uneventful; however, inter-

Airmail delivery of a needed part.

esting to me, it was pretty much repetitious of the down trip.

Wednesday was change day when the new crew came on and, just outside the Federal Lock at Troy, we were nailed by fog again. It was discouraging to sit and wait when we were only 5.87 miles from Lock 1 where I would be getting off. Then my trip on the ERIN was over. It was 7:15 when I said farewell to my friends, jumped into my car, and headed for Spencerport and home.

CHAPTER 6

ON THE ERIN KEHOE – 1980

I used every means of public transportation save one, to make this particular trip in late August of 1980. I drove to Albany from my home in Spencerport to do some research at the New York State Department of Transportation. As usual, it was good to see my friend, Sy Foltman, again. Sy has been my fountain of knowledge of state records about the canal. Once when Sy wasn't there, I asked his boss, Joe Stellato, who took Sy's place when he was absent. Joe almost blasted my head off as he yelled, "No one can replace Sy Foltman!"

After my morning with Sy, I called Marty Kehoe to see if I could catch a ride on the ERIN as I had in past years. He stunned me by saying that the ERIN was tied up for lack of work. It never entered my mind that any of his boats were ever out of work during the canal season. He offered to let me ride on either the JAMES or the MARTIN, but I said no thanks. I wanted to be on the ERIN with the men who had grown to be my friends over the past four years, especially her Captain, Bob Gordon. I told him I would hang around for a few days to see if he could turn up some work for the ERIN. I knew Marty would not merely sit and wait. He would no doubt be "burning up the telephone" trying to drum up some business. The waiting was the longest, slowest, more boring time I have ever spent.

Finally by Monday, Marty had a job but was worried about rounding up a full crew. I was to call Bob Gordon at his home in Whitehall, New York, and make arrangements to get to Staten Island. The ERIN was tied up in Carl Eklof's yard. Eklof and Marty are partners, but not in the normal way. It is not a corporation with each of them an officer. Marty owns the tugs and Eklof owns the oil barges. They work together on the bids to haul liquid cargo.

After an hour and endless busy signals, I reached Bob. He told me what train to catch in Rensselaer (across the Hudson from Albany) and that he would look for me at the station. Bob and the mate, Jimmy Day, would board at Whitehall and the rest of the crew at various places along the way. The train was to arrive at 5:30 p.m., so I had plenty of time to make arrangements to leave my car at Lock 1 in Waterford. The taxi from Lock 1 to the station was $14.00 for the twelve miles. For only $15.75 I rode the train for 150 miles. I hadn't been on a train since World War II. Both the train and the station were new to me.

Arriving at the station an hour early, I found it to be small, neat, sterile, and sprinkled with cheap molded plastic seats – exactly 123 of them. I had plenty of time to do such constructive things as counting seats and waiting passengers. The station was a far cry from the vaulted ceiling, granite buildings of the past. There was a certain grandeur about them that appealed to me. I made the mistake of not grabbing something to eat in the station. I thought I would wait and eat on the train. Wrong! When it arrived, the train was packed, with every seat taken. I had heard stories of passengers standing for the entire trip. I wasn't looking forward to that possibility. After about an hour, another section was added.

Bob and I did not find each other at the station. Panic set in. If the crew missed this train, could I find my way to Ecklof's yard? I gave up trying to fight my way to the food car. One thing hadn't changed in forty years. The train stopped at every little town and took over three hours to get to New York City. I had made up my mind that if Bob was not aboard I would get a hotel room because I was not about to travel alone to the waterfront on Staten Island late at night. It was a great relief to spot Bob and Jimmy at the head of the ramp. We waited for the train to empty because we still had to pick up one of the engineers, Jack Russell, and the cook (for obvious reasons, this cook shall remain anonymous). The cook was the last one off the train, and his weaving walk brought an instant look of disgust on the Captain's face. He said, "We're not waiting for the #%#@%&%%, let him find his own way to the boat." All of us were hungry, so we stopped long enough to eat a bite at one of those super fast, frantic paced, food dumps that are typical in New York City. It's a wonder all their customers don't perish from indigestion.

We now headed for the subway for the ride to the Ferry Terminal. Another shock greeted me. It was 1941 when I last rode the subway. I was not prepared for the graffiti that covered every inch of the cars, both inside and out, plus the random overlay of paint left by some spray can maniac. All the horror stories about riding the subway came rushing into my head. I was very glad to be traveling with four other men.

We reached the Ferry Terminal at 10:40 and only had a few minutes to wait. To our amazement, the cook had beat us there without getting mugged. Once on board, we moved as far away as possible from the cook. The eighteen-minute ride to Staten Island was the best part of my trip, so far. It was a beautiful balmy night, but I was a bit annoyed that the natives aboard ignored the lighted Statue of Liberty. I realize they are rather

Tug didn't slow down enough. Result, man pasted to girders.

Hanging on for dear life.

blasé about their surroundings, but I don't see how anyone could tire of looking at that grand old lady. The last stretch of my journey on the bus is difficult to describe. I witnessed a mechanical miracle. It was a bone busting punishing experience. I don't see how any machine could take that much abuse and still function. The level of noise was ear crushing. Everything that could rattle, squeak, grind or bang did. I doubt if Richmond Terrace has been repaired in twenty years. Conversation was impossible. This could have been a modern version of the "One Horse Shay," we read in grammar school. The driver's lead foot on the gas pedal didn't help any. Ecklof's yard was a welcome sight. We were tired, hungry, dirty, and irritable. One trip was more than enough for me, but these crewmen go through this all day agony quite often. This has helped to dilute the romantic notions I've always had of their way of life.

Cook got the coffee pot on with impressive speed. This was one way he hoped to redeem himself with the Captain. I had a flash for him. It was going to take a hell of a lot more than a cup of coffee to do that. Bob has a very low threshold of tolerance for people who travel and board his tug with a load of booze under their belt.

Otto Andereasen, old friend and deckhand, was there to greet us. To keep us from starving, Otto was kind enough to bring us some donuts. Otto lives in Brooklyn and is big enough to take care of himself, yet he traveled to the boat in daylight. On crew change day, the men will not get off the tug in New York City at night if they are alone.

I hadn't seen Otto in two years and was looking forward to renewing our friendship. He is a self-effacing man born and raised in Norway. Everyone jokes about these "square heads" as the boatmen from the Scandinavian countries are referred to. I'm part Dane so I've heard all the stories. The usual ones are: Don't force it, get another square head, or because he is rather rigid in his work habits, they call him Otto-Matic. They joke with and about him but he has the respect of everyone. He is a top notch deckhand and more like him would be welcomed by any towing company. He was a fisherman as a young man during World War II and would not talk about life under the Nazi occupation of his country. I respected his privacy and did not press him for details.

Everyone was in the sack by 12:30 and up by 5:30 a.m. We had no orders yet to move. The crew was ticked off. Dave Kehoe, the other deckhand, like the rest of the crew, had just gotten home (Vermont), when he had to turn around to come back to work. The crew doesn't like to be jerked around like that. It's strictly a matter of economy. In the old days, 1930s and 40s, some owners would let a crew sit idle for a few days before paying them off.

We were in with two of Ecklof's tugs and the MARTIN KEHOE, so work was really scarce. It was 7:00 a.m. and we were still waiting. The sun broke over the shed on the dock, telegraphing the news that it was going to be a beautiful but a very hot day. At 8:30 the yard superintendent asked Bob to move the MARTIN and ourselves so he can shove the E-23 ahead in order to weld some holes in the hull. This is the barge we were waiting for.

One door to the galley was still locked, and no one knew where the key was. In desperation, Jack Russell used a sledge hammer and cold chisel to beat the hell out of the padlock. Not being a dime store lock, it took some time before it shattered. We had to do it because the galley was like a blast furnace. As we waited for work, the daily routine continued with Dave painting. No one could locate Dave Oliver, so we were short one engineer.

Any kind of job would have been welcome, as the heat was getting to everyone. Hooray!! Got a job at 10:30. We went out to Stapleton anchorage to pick up an empty barge that had just finished refueling a ship. It was a short run but the cooling breeze gave us little relief from the heat. Dropped the barge off at Belcher's dock and back to Ecklof's by 11:45. Our timing was lousy. We should have slowed down so we could have eaten our full turkey dinner out in the harbor. It was just too damned hot in the galley to enjoy it.

After lunch we were still waiting. A tiny breeze came through the wheelhouse but there was none on the deck, for the barge and sheds blocked any breeze. When it's this hot, the cook shuts down the range as soon as the food is cooked. If he needs to reheat any of the food, he uses a hot plate. Normally, a galley range is never shut down. It is oil-fired and burns constantly and operates entirely different than any type you may have in your home. I couldn't get used to how much chow Jimmy Day can put away. After eating two helpings of everything, he polished off a candy bar and a bottle of orange pop.

At 3:30 p.m. we had to move again. Ecklof wanted to lay a barge in where we were tied up, so we just shifted to the other side of the pier. We barely got our lines out and we got another job. This time we are going to move a derrick boat from Hess Bayonne to Greenville, N.J. There are no deckhands aboard the barge, so Otto would ride it to handle lines. Because this is not his normal job, Otto received extra pay for doing this. I kept wondering why the darn thing didn't tip over. It certainly was top heavy and awkward to tow. It was "no big deal" for Bob or Jimmy. They have to be able to handle anything, and they do. We finished up this job at 6:00 p.m. What started out as a lost day for Marty Kehoe was turning into a profitable one.

Our last job of the day, one that would end after midnight, was swapping an empty scrap iron scow for a full one. Jimmy started for 4th Street in Port Newark instead of 4th Street Brooklyn. After this was brought to his attention, he did a quick 180-degree turn in the middle of the harbor and headed for Brooklyn. Naturally, everyone in the crew had to give Jimmy a "shot" over that boo-boo. Arriving at the dock, Jimmy turned the wheel over to Bob so that he could help Dave handle lines. It's not normal for a mate to handle lines, but they didn't want to get Otto out of the sack for the short time it would take to do this chore. I have seen this crew many time show this kind of consideration. This job was dangerous because loose scrap was all over the narrow ledge around the scow which made the footing treacherous. To make matters worse, there were no safety cables to grab hold of. Dave and Jimmy had to loop their arms over the sides of the scow and inch their way along. This kind of work puts an added strain on the Captain who is always concerned with the safety of his men. When they had finally reached the dock, there was another problem. They couldn't untie the lines holding the scow to the dock. The men who had loaded the scow had piled a lot of scrap on top of them. This had been done to prevent the neighborhood kids from throwing the lines off and letting the scow drift away. Just another example of what a zoo New York City can be. The crew had no choice. We passed a fire axe to them to chop the lines off. We could have used a more powerful tug, for as it was we barely made any headway against the flood tide.

Wednesday we were up again at 5:30. It can be a surprise each morning as you wake up in a different place. This time we were at the Chevron dock in Perth Amboy, ready to be underway soon. Engineer Bill Spring came aboard last night after I had hit the sack.

Yesterday had been a special day for me. I always wanted to see what harbor work was like. It takes impressive skill for any tug crew to navigate through all the traffic and confusing lights and find a pier they have never been to. Ecklof changed his mind, so now we are going to take the E-17 instead of the E-23. We finished making up to the E-17 about 7:30. Neither Bob or Jimmy like this barge. The crew's house was built on top of the deck instead of in the hull as most are. This means they will have to use the upper wheelhouse for the entire trip, which is not as comfortable as the lower wheelhouse. They can't see over those houses even when the barge is loaded.

This was my first trip in really hot weather, so it took a while to get used to seeing Otto and Dave wearing shorts as they worked. We cleared the Statue of Liberty on a hot hazy day at 11:00 a.m. We had a load of fuel oil for Green Island, New York (across the Hudson from Troy). Since we were loaded to 13 feet, we needed three hours of flood tide in order not to ground as we unload.

One o'clock on the river and the traffic was nil. One trap rock tow coming down. At supper time we passed West Point, and again the day was so hot the cook had to shut down the range the minute the food was cooked. At home, you'd probably have a light salad on a day this hot. Not so on a tug. Three big meals a day, everyday. Dave is doing a painting job that he is not required to do. Someone had painted the name ERINKEHOE in one string. It was very difficult to read, and looked terrible. We hope Marty appreciates his efforts.

I met another of the rugged individuals whose looks fool you. He was Bill Spring, our engineer from Castleton, New York. Tough looking and tough talking, with a cigar always jammed in his mouth, never removing it when he talks. He reminds me of the late movie star Edward G. Robinson. Spring is an officer in the National Guard, and I can picture him terrifying some green lieutenants in his outfit. Judging by my own experiences in the Army and Navy, most young officers need someone like Bill to shape them up. First impressions can fool you. Beneath that rough exterior, I found a very competent engineer, a devoted family man, and someone I'd like as a friend.

Bill would like to see his son have a chance to become an engineer, but he is up against a Catch-22 situation. The towing companies only hire experienced engineers, so there is no place for a young person to learn the business. The entry level job used to be that of oiler, an occupation that has been eliminated. Decking is the entry for the wheelhouse. There should be a joint sponsored apprentice program funded by Local 333 of the Marine Union and the owners.

At 5:15 a.m. we passed Ravena, New York. Every time I went by there, I would think of my friend Captain Jack Maloney and the tug JOAN TURECAMO. I've been with him when he came here to pick up the cement ships and take them to the turning basin at Albany. They are too long to turn around at Ravena. It's foggy and cool, and good sleeping that night. Jimmy was on watch and he either had the radio or his tape recorder on, listening and singing along with his favorite country-western music. Jimmy has excellent eyesight and what I would call "river sense." He seemed to know where each buoy was before we got to them, and sometimes he used the searchlights to pick them out of the mist and fog.

Cook put our remaining food in the freezer because the reefer [slang for refrigerator] is acting up again. We were also low on food, but we knew there'd be time enough in Troy to grub up and get the reefer fixed. Bob was gone most of the day, for he had to testify in court because of a suit filed against this tug as a result of ramming the gates of the Federal Lock at Troy.

As we approached the new expressway bridge, Bob pointed out an oddity. The roadway lights extend from the Albany side to the middle and then stop. From there to the east side, there are no lights – the result of a political fight over them. Weird, very weird.

We reached Green Island by 8:45 a.m. This was the least adequate set up for pumping out I've ever seen. No dock, no pier, no dolphins, no nothing. Like monkeys in the wild, the employees of the oil terminal literally crawled through the jungle of underbrush to locate a hidden cable that we could tie up to. There was a "Mickey Mouse" pipe frame to which the hoses could be attached, but it wasn't strong enough for a canoe to tie up, let alone a tug. Cheap is one thing, but this was ridiculous.

As soon as the barge was secured, we broke away and went over to the Troy wall to settle in until the barge was pumped out. Pumping would take about six hours. Since were tied up just a block from the business district, I decided to wander around in order to see the changes that had taken place since I was a kid. I was also looking for a fruit store and a bakery. I found just what I wanted and bought some baked goods for the crew and a dozen bananas for Otto. I remembered from my first trip how he used to ask the cook to bring him bananas. The crew always kids Otto that bananas are just the thing to strengthen one's virility. Otto always would blush a little, but he never replied to their friendly taunting.

The weather warmed up slightly, and it sure felt good after the intense heat of New York harbor the day before. The barge was empty of its 20,000 gallons of #2 oil by 3:00 p.m. After taking the barge alongside, we headed down river to Cirello's dock at Albany to fill up for our trip north.

While the barge was being filled, we headed for the Mobil dock to take on fuel and water for the tug. As long as no one was waiting to use the dock, we could stay there until the barge was ready. It was a lovely warm evening as the men gathered on the bow engaging in their pastime – swapping lies and telling tall stories. We left the CIBRO dock at 11:00 p.m.

Friday we woke up to a cool, cloudy day, which stayed that way until noon, when the temperature shot up like a rocket. I had a new experience as a result of taking this trip in hot weather. Sitting at the bottom of Lock 8 turned out to be worse than Dante's Inferno. Luckily for us, we usually spent less than 15 minutes in this position.

Rarely did we have a really lousy meal, but as I looked over my notes for this day, I found I had written just three words, "Supper – fish, Ugh!!" As a matter of fact, the guy may not have been the world's worst cook, but I think he could incite a riot with his cuisine.

After supper, rumors were flying all over the boat, rumors that the boat would be tied up again after this run. Bob said no, but we switched barges in the Hudson on the way down. As I headed for the sack tonight, various things came to mind. For example, I always remember to step over the coamings, but forget to duck my head. Door openings on a tug are not as big as those at home, and I've found that it takes a few days after one of these tugboat trips before my head is fully healed.

Another difference between the tug and home has to do with going to the bathroom (head) at night. While this subject is usually not discussed in books, I feel it's important to provide a full picture of what life is like aboard these old tugs.

Except for younger members, the rest of the crew have to get up at least once a night. Cutting down on liquids after supper doesn't really help solve that problem. At home, it becomes a conditioned response, often accomplished in a trance-like state, with the bathroom close and warm. It takes so little time we don't ever become fully awake. Aboard a tug, however, it's a whole different ballgame. You lay there fighting a losing battle, hoping stupidly the need to go will leave. But it doesn't go away, so – alert but not necessarily too coordinated – you stumble around trying desperately not to wake anyone else. Groping around for shoes or slippers for your feet, you may say to hell with it and head for those cold, clammy steel stairs. You try to remember to duck your head as you open the sticky door as quietly as a cat burglar. Out on the deck and over the side. Not too bad on a warm, dry night, but try to imagine this exercise on a cold, rainy night. It's murder, just plain murder.

Leaving Whitehall, we ballasted down the bow of the tug. The trim of a tug is controlled by the amount of water in the bow and stern tanks; with the stern down, the tug has more pushing power but obviously rides deeper in the water. The marshes north of Whitehall are very shallow; therefore, we have to sacrifice some speed until we clear them. With the deeper Lake Champlain water, Bob can put the stern down as far as he deems necessary.

By 8:00 a.m. Saturday, we made Plattsburg. It took us three days to go from the Statue of Liberty to Plattsburg, New York, and was a quiet morning as Dave polished brass in the wheelhouse while listening to Jimmy's radio. It was a strange day, too. Although it was morning, the sky looked like sunset because of the haze. I had the feeling there was foul weather not far away in our future.

Every place we tie up seems to be different, and this place was no exception. There are three dolphins (to tie to) out from shore but no walkway to get to them. There is a small platform on top of one of them for the terminal crew to hook up our hoses to pump out our barge. With no walkway, Bob had to run the tug to shore to pick up the men from the terminal. By 9:30 both deckhands were having a "field day" in their quarters. Field day in G.I. language is cleaning your bedroom. It's done a little differently aboard the ERIN. As in the service, mattresses are brought on deck to be aired out. The best way to do the next part would be with a vacuum cleaner, which of course we don't have. So we use the next best thing, an air hose. I personally think they did a fantastic job of redistributing the dust and dirt. But turning Otto loose with an air hose is like handing a pyromaniac a book of matches.

It was three o'clock in the afternoon with a stiff breeze blowing. To break the monotony of waiting for the barge to be pumped out, I walked up to the main road. It wasn't worth the effort. Just that short distance from the lake and the temperature and humidity soared. Oh yes, tug crews spend a lot of time sitting and waiting.

The 3:00 estimate for pump out was off by two hours. We made it up to the barge and were headed down the lake by 6:30. Within minutes, we were enveloped in an incredibly violent storm. Thunder and lightning and rain as thick as any I used to see in the South Pacific shortly after World War II. As we were the only vessel on the lake, I began to wonder if we were tempting fate by being out here. The crew takes it in stride, having been through storms like this many, many times. It was like a million flash bulbs, and the next second – total black. The lightning bounced off the opposing mountains like a rubber ball gone berserk. The thought crossed my mind that there would be some folks nearby living by candlelight tonight. Little did I know it would also affect us. From the unobstructed view in the middle of the lake, it was a beautiful, if slightly terrifying, sight that few people have had the opportunity to witness. The intensity of the thunder was magnified by slamming against the Green Mountains of Vermont, ripping across the lake to be bounced back by rebounding off the Adirondack Mountains of New York. It is a humbling experience to be reminded of the power of Mother Nature.

Sunday at 6:00 a.m. we were tied to the wall just south of Lock 12. It was too foggy to move. Usually when we are stopped for fog, I tried to write some observations about the men. This time I wrote about how they remind me of old war vets (which some of them are), and yet age is not really a factor. Whenever boatmen meet other members of the fraternity or even amongst themselves on the ERIN, they talk of other boatmen, past and present tugs they have been on, and places they've been. They never seem to tire of doing this. Another facet of this phenomenon is the bravado shown as they try to top one another, on what a splendid job they did, telling off the owners, dispatchers, etc., not to mention their finely honed skills of throwing punches and emptying bottles of booze.

The conversations are not all in this vein. I enjoy it when they talk about the men they have admired, who have passed on long ago. Also the humorous stories they tell and the practical jokes they have witnessed or perpetrated. Like war vets, I think part of it lies in their partial isolation from what we might call the "normal" life that most of us live. Their skills are unique and their numbers are dwindling rapidly, and this tends to knit them closer together. Their off hours each day are spent together as opposed to the rest of us who head for home after the work day is finished. Most of us probably do participate in "bull sessions," but certainly never to the extent boatmen do.

The fog lifted by 8:00 and we were anxious to get started. There was no need to hurry because we discovered Lock 11 had no power. The storm the day before had knocked out the main transformer. We couldn't leave until noon. While we were waiting, Bob tried to call the boss, but all he got was Marty's telephone recording to leave a message. Yesterday he and Jimmy both tried all day, with no luck. Getting that recorded message drives them both "up a tree." Since we can't reach Marty, I'm going to get off at Lock 1. I don't want to get hung up in New York if they have to tie up the boat again.

I'm sitting alone on the stern plates as I write this. At certain times, and this is one of them, the vibrations on the boat are just right to make these plates rattle with intense noise. Before the noise drove me off, I spotted

Bill Spring in his normal steaming position – cigar in his jaw and one foot on the rail, staring off into space. The cook may get off at Lock 4 as we are almost out of grub. This cook was strong on pudding, jello and packaged cakes. I never did see a pie on this trip. No fried ham or hash browns either. He missed the tugboat cooking standards by a very large margin.

I had to duck inside quickly as a heavy shower hit us. We buttoned up everything on the starboard side. Barely finished, when it stopped and the sun popped out. As these summer storms often do, it left the air cool and clean smelling. The tug and barge glistened where the sun struck the raindrops that cling to everything.

The fog caught us again a few minutes after midnight. We can't use the searchlights because they blind you. It also happens when you turn your high beams on, while driving in a fog. We could have run a lot longer if the buoys had lights on them as they do on federal waters. This is another cost saving move by New York State. The buoys are nothing more than 55 gallon oil drums with a dime store reflector on them. This could very well be the last night I ever spend on a tug. I don't like to dwell on it because it depresses me, probably more than it should. We reached Lock 1 at 7:30, and I said my farewells to all. I hung around until they were out of sight.

CHAPTER 7

CAPTAIN JACK MALONEY

I first met Jack Maloney aboard the tug JOAN TURECAMO (formerly the MATTON 25) on September 13, 1974. As the Albany harbor tug, she was tied up in her usual spot on the wall of the old lumber dock on the Rensselaer side of the Hudson River. She earns her keep by docking and undocking the ocean-going vessels that come to Albany.

My first impression was how much Jack looked like my Dad. The same build, height, and weight, with a real Irish twinkle in his eyes. They could have passed as brothers as far as looks go, but there the resemblance definitely ended. Their ways of life were miles apart. Jack spent a lifetime of adventure and travel in the towing business, while my Dad stood at cutting tables in the shirt factories of Troy, New York, all his life.

Jack was born in Cohoes, New York, on January 25, 1910. After attending St. Agnes School in Cohoes and LaSalle Institute in Troy, he got a job sweeping floors in the now defunct Harmony Mills in Cohoes, for $11.00 a week. For many years, these mills were the largest of their type in the world. From the depressing sameness of the mill-owned houses to the drab mill buildings themselves, their presence, both physically and economically, dominated the city of Cohoes. It was expected that Jack would follow his Dad's footsteps into the mill, but Jack was saved from the life of a mill hand a year later when his brother-in-law, Frank Bertrand, got him a deckhand's job on the tug EMPIRE.

Working in the fresh air, with lots of good food and a place to sleep, was the icing on the cake; besides, he was earning over twice as much as he did as a mill drudge. He spent the first year learning his trade as the tug EMPIRE worked New York Harbor and the Hudson River.

The following year, Jack went decking on the P.C. RONAN for the John E. Matton Co. and stayed with Matton until Matton sold out to Turecamo in 1964. After that, Jack worked for Turecamo until his retirement in 1976.

Jack's learning really began on the P.C. RONAN under Captain George Van Steenberg in 1929. Some might say George Van Steenberg was lazy and irresponsible; and that may be true, but Jack benefited from his life style. The faint of heart didn't last long under George. I asked Jack why. Jack responded, "He was a great man to learn with. He'd say, 'Take the wheel.' In fact, the first couple of weeks, when I was on the P.C. RONAN, four or five lads were playing cards in the pilot house and we got an order to go to one of those dredges or something, and he said, 'Jack, you take the wheel.' Because there was only one entrance to the wheelhouse, I had to climb in the window. I then steered the boat down by the dredge, landed alongside it very easily, climbed out the window, and got the lines out. And this was the first time I was on the boat."

Q. "So you did the whole thing yourself – lines out – steer it?"

A. "Eager beaver. Anything they wanted, I would try it."

Jack finished out the season on an old steam tug, the WM. R. Hulsizer. The following spring, George asked Jack if he would like to come back to the RONAN. Jack was more than happy to get off the dirty steamer to a nice clean diesel.

I asked him how the season started.

A. "We picked up an oil barge and headed for Buffalo. We only had to run days, but we'd run till midnight and tie up. The next morning, I would wake him up about 5 o'clock to start out. He'd say, 'Oh, you take it out.' Well, I'd never been up the canal before. I didn't even know what an oil barge was to push. He'd say, 'You take her out.' His opinion was that you could do nothing wrong. You're in the canal and you can't get out."

Q. "Couldn't you run into the bank or something?"

A. "Oh sure, the canal was narrow in some places and when you would meet someone, you'd slow down, your slowest speed until you got by them. Well, lots of times . . . I would put it on the bottom too hard and I'd come up and say, 'George, I'm on the bottom here.' He'd say, 'Oh, to hell with it, we don't own a rivet in 'er. We'll get it off.' That's the way he was – jovial. He'd get out of the bunk and get her off and we'd go on.'

"The Captain used to bring his wife on board for a few trips. Having a woman aboard is not that unusual, but when it's the Captain's wife and the boat is only 52 feet long, it certainly makes life a little more difficult." Jack recalled a couple of examples. "The Captain's room was directly behind the pilot house with only a thin partition between. Sometimes the Captain and his lady would retire for an afternoon siesta. During these times, Jack would have to modify his normal steaming position, which was sitting on a stool and resting his head against the bulkhead. Sometimes he would find himself bouncing off the bulkhead from the vibrations that did not originate from the engine room. He would have to maintain an upright position until the "siesta" was over for the day.

In the 1920's and 1930's, some tugs didn't have toilets. A bucket placed in the engine room was used. This was true on the P.C. RONAN. It was disconcerting, to say the least, when Jack was in position on the bucket and the Captain's wife would pass by the door and yell down a greeting to him.

I asked Jack about pay and promotions. The pay for a deckhand was $3.00 for a twelve-hour day. On a day boat, if you worked over twelve hours, you received $.50 an hour extra – Saturday and Sunday were straight time. The first raise they got was 46¢ per day more when the union came in 1936. Jack spent three years as a deckhand. Because of slack work, he spent one year as a cook. The next year, he steered as a mate. 1938 was an important year. He became Captain on the tug JOHN A. BECKER, and married Dorothy Gentile. I should have mentioned that the $3.00 a day may have sounded like a small amount of money, but there were some companies, such as Standard Oil, that paid only $55.00 per month ($1.83 a day).

I asked Jack about the requirements for his license.

A. "Three years as a deckhand, and you can only take the test for the waters you have traveled on, and you need a letter from the Captain you worked for. Now you need twelve trips plus letters from the Captains and companies. Your license has to be renewed every five years. The only examination at that time is an eye test.

Jack's first year as Captain was spent hauling molasses to Buffalo. The trip took 26 days, traveling New York to New York. I asked why it took so long. He said, "Underpowered tugs (150 to 350 HP), current on the 60-mile level [between Rochester and Lockport], and bad design of the molasses barges." I was surprised that there would be an appreciable amount of current in the canal. It took them 72 hours to run the 60 miles. This was slow enough for the off watch to step off the top of the tug onto the bridge at Brockport, go to a movie, then walk to catch up with the tow after the show. It was also the last year Jack had to find a job in the winter after the canal closed. He started working in New York because the war in Europe had put high demands for goods to be shipped overseas. It was the first year Ralph Matton had steel tugs to send down. Winter jobs were hard to find for many boatmen because if the employer knew you worked the boats, he was reluctant to hire you because you disappeared in the spring. Before we left this decade in Jack's life, I asked him if he would talk some more about the depression years.

My first question centered around the possibility of men resenting the fact that he had a steady job when they had none.

A. "If there was any resentment, I didn't notice it. We'd come into Waterford with a tow, and there would be 25 to 30 men on the dock. Some of them were Captains and mates who couldn't get jobs, and they had all kinds of experience. They would beg for even a day's work. It wasn't all beer and skittles. There were slack times, but we were very lucky working for Matton. There was little work in 1934-1935. Sometimes five or six boats would be in the yard, but no one got laid off. Matton put them on half pay. They kept as busy as they could putting the tugs in tip-top shape. When that was done, we would put on their bathing suits and go for a swim. After the swim, we'd have supper and go home."

Q. "Many people had a low opinion of boatmen. Do you think it was justified?"

A. "The boatmen were looked upon as bums because they were away all summer. They might be in Tonawanda or someplace else once a month. They didn't know anybody, so they'd go to a saloon for a few drinks and catch up on the latest news. They might get drunk, therefore, everyone thought that happened at every stop and that they were all drunks. They didn't stop to think that those not on the boats could do it every night and most people wouldn't think anything of it."

Q. "Yesterday you spoke of making your own entertainment in the 1930's and 1940's. Could you give me some examples?"

A. "When I was on the JOHN A. BECKER, we were pushing Matton's oil barge, JEMSON 1, and it had a big flat house [crew's quarters] up front. Two young Polish lads from Amsterdam were on the barge. In the winter, they played the accordion for country dances. One of the deckhands played the mouth organ, another played drums, and the old cook – Harry Powers – was a pretty good man on the banjo and mandolin. After supper, they would set up on the roof of the barge house. They would play as we went along and the rest would be singing with them. We'd be going through New York harbor and the ferry boats would be going by. The people on board would rush to the side as we passed and listen to the band. If we went by the office (Matton's), everybody would drop what they were doing to see the band going by."

Q. "About what year was that?"

A. "1940 or 1941. We were going up the western canal and there would be a line of cars following right along in the evening. In Lockport, they would dance in the streets. We had simple pleasures then. We had an old hand-crank ice cream maker, and a clam steamer so we could have a clam bake in the fall."

Q. "What are the differences in grubbing up, now and then?"

A. "When I started out in 1929, we got 90¢ a day per man for food. Out of that you had to buy the Captain's cigarettes and tobacco. We even had to buy soap and toilet paper for the boat. In the early days with ice boxes, we could only go about three days before we had to buy food and ice. I remember when I was on the JOHN A. BECKER as a deckhand. It was the last tug Matton had with an ice box. We sometimes had to steal the ice. We'd get into Waterford at three or four o'clock in the morning, and there were a couple of stores that would open for you, but not the ice houses. There was an ice pond in Cohoes where they used to harvest ice. We'd go down ourselves, unlock the door, and steal the ice. We did it because Matton only allowed three hours to grub up, get ice, coal and water. He didn't care how you got it. In Syracuse, we needed ¼ ton of hard coal for the galley range, but they weren't interested in selling only ¼ ton, so we stole that, too."

"Some of the stores would throw in some cigarettes and maybe some strawberries when they were in season. Some would pick the cook up and bring him back to the boat. No more of those things today. You buy things in a supermarket, and they could care less if you did business with them or not. When I worked in New York, all you did was call the dispatcher and tell him you needed grub for seven days and the provision company delivered to the boat. You didn't have to tell them what you wanted. Getting refrigerators and freezers made this change possible."

Q. "As long as we are on the subject of food, did you ever buy from farmers?"

A. "We did in the Depression days when we towed the wood barges with families on them in the Champlain Canal. We didn't run after dark, but if you got caught between locks, you didn't try to make the lock. Instead, you would tie to shore. There was usually a farm nearby where we bought fresh milk and vegetables. Sometimes we would even go to a country dance. We did do something else when we were running the western canal. On the last trip to Buffalo, we would stop at Lock 25 (May's Point) at a little store that catered to the boatmen. We would order barrels of apples, potatoes, squash, and gallons of maple syrup. On our way back, we would stop and pick it all up. Also, on the last trip of the season there was a little more drinking than usual."

Q. "Again on the subject of food, can you remember a really great cook?"

A. "I sure can. John E. Matton had almost all Portuguese cooks before he died. We had one named Julio Gomez, and everybody complained that they'd get too fat because he fed us so much. The table would be filled wall to wall. If he had spaghetti and meatballs today, he'd have a big roast beef, mashed potatoes, and all the trimmings the next. On Fridays, he had three kinds of fish for the Catholics – one was always scallops. The table was always covered with pastry. When we went to Clarkson, Canada, there was a farm nearby. I would walk there with Julio to keep him company. He would buy fruit and vegetables by the bushel. A crate of strawberries went for 15 a quart; peaches a dollar a bushel. Julio's family lived in the Cape Verde Islands, so he had no place to go. He never got off the boat for three years. He was on board to save his money to bring his family to the United States."

Jack said if the cooks had any money left over, it was theirs. I got an obvious answer to the question, "Did anyone take advantage of this?" "Of sure," he replied. "I can assure you it goes on today, too."

Q. "Did you know anyone with odd eating habits?"

A. "Sure, Jimmy Clinton. I'd have to make a phone call and he'd say, 'Get me a dollar's worth of candy bars,' and that was when they were a nickel each. He would sit there and read and eat the 20 candy bars, or he'd say, 'Bring me a quart of ice cream,' and then Jimmy would devour that, too."

Q. "Was he a big man?"

A. "He could lose 60 pounds and put it right back on whenever he wanted to. He'd come off watch at midnight and eat a whole box of Ritz Crackers with peanut butter and jelly before he went to bed. Or he would take a box of corn flakes and a can of fruit salad and finish that off in a few minutes."

Q. "As long as we are on the subject of odd things, how about odd behavior?"

A. "One quickly comes to mind because we are near the place where it used to happen – on the Hudson near Rensselaer. I had an oiler on the MATTON 20 who lived in Rensselaer. It was during the days when we worked from spring to fall with no time off, except what you could steal. He'd want to go home for a bit, so he'd strip off his clothes, dive into the quick water, and swim to shore. I've seen him do this on a dark, cold, rainy night. His wife would pick him up and he'd catch the boat up the canal the next day."

Q. "With all the time you spent on the canal, you must have gotten to know a lot of the lock tenders."

A. "We knew practically all of them. When we were towing the old wood canal boats, we would have to double-lock. In other words, you had to lock through the first two boats, tie up, and wait for the men to winch through the next two with the power winch that was on each lock wall.

You had an hour to yourself to chew the fat. Most of the locks sold something – ice cream, candy, smokes. A few locks had farmer's wives who came down with homemade bread, doughnuts, and cakes. There were a half dozen locks where the farmers came down to sell surplus milk."

Q. "Can you think of some tows that were a little out of the ordinary?"

A. "Right after I went Captain in 1940, the first two Navy Mosquito Torpedo boats came down the Great Lakes to Oswego, New York. They sent us up to get them because their engines were so powerful that they couldn't throttle them down enough to run in the canal. They would have washed the banks away. I bought a big timber in town, strapped it across the stern of the two boats so they could run alongside each other, and I got behind with the JOHN A. BECKER and pushed them to Troy.

"We ran just days. The boats carried a full crew of officers and men, but their quarters weren't set up. We'd stop about four in the afternoon so they could get to a restaurant and a hotel for the night. One of the officers was nice enough to give me the keys to the liquor and told me to treat the boys when I felt like it."

Q. "Can you think of any firsts while you were with Matton?"

A. "I think I was the first one to push wooden canal boats, in the same fashion we now push oil barges. We were running into the Solvay Plant at Syracuse for soda-ash (boatmen always called it "sody-ash") for Connors Marine. We were using the old narrow wood canal boats (22 ft. × 100 ft.) and towing four at a time. We were loading a ship at Coxsackie on the Hudson, so I got the idea of lashing them together and pushing them. They were narrow enough so that two side by side would fit the locks. Pushing is faster than towing because the quick water from your propeller hits the front of the barge and slows it down."

Q. "We've talked about a lot of things that were odd or strange. Let's try one more. How about a weird place to take a tow?"

A. "We had a few trips to St. Albans, Vermont, about 1937 or 1938. To get there, you had to go past Plattsburg and swing west between North and South Hero Islands on Lake Champlain. It was called the "Gut" and was very shallow. There were two little railroad bridges with a family at each bridge. They were both old couples who would come out with this huge key and manually open the bridge. The man was on one end and the woman on the other. They would turn it like you were winding a clock. After we cleared the bridges, we would head up St. Albans Bay with these Socony Oil barges. It was so shallow we could only load eight feet. Once through the "Gut," we had deep water. One thing about the "Gut" is that it's filled with weeds that were sometimes as high as the barge. It was a strange feeling, but we pushed right through."

Q. "Your story about the "Gut" leads me to my next question. What are the easiest and toughest places to work?"

A. "I think the canals are the most difficult because the channels are all pretty narrow. If you weren't familiar with them and the lights were out, you could easily go aground – which many have done. You had to operate so that the upgoing outfit had to lay right up against the bank. There was always the chance of putting a hole in the barge. You had to give the guy coming down all the room possible. I learned something different each trip I made."

Q. "What is the easiest place to operate?"

A. "The Great Lakes, because there is a lot of open water and they have up and down-bound traffic. The ships have lanes going up and down a couple of miles apart. Before radar, they had marine phones telling what ship was coming from Buffalo and steering a certain compass course. You knew where they were and where they were going."

Q. "Do you have any stories about working on the Great Lakes?"

A. "Sure. What comes to mind was a job to tow four dead ships from Milwaukee to Quebec. These ships would be wild coming down the lake. We would be towing them, one at a time, on a 1,200 foot hawser (pronounced hah-sir), and they'd run so close that sometimes they would be alongside us and a ship would pass and they'd call back and ask if we were towing it or was it running free, or what was happening. You would have to tell them that we were towing it, but you'll have to give us plenty of room, she's a wild one.

Q. "Wouldn't it have been easier to tow alongside?"

A. "No. It was much too rough and too much of a problem to get your lines up."

Q. "How long did it take and how large were the ships?"

A. "They were 600 feet and drew 18 feet of water, and we had four men riding to handle lines. Then we would pick up four more men at Port Coburn to help handle lines in the Welland Canal. We had to use the extra men because there was no power on the ship to operate winches. It was pure bull work. We picked up a tug at Cape Vincent to help us down the St. Lawrence River and through the Seaway locks."

Q. "Did you have to take on a pilot?"

A. "We took on two pilots and paid for three."

Q. "I don't understand."

Capt. Jack Maloney in wheelhouse of JOAN TURECAMO (was MATTON 25), Albany Harbor, 1974.

Capt. Jack Maloney's musical crew.

A. "We had a pilot on each tug and one for the ship who wasn't there. This was Canadian rules, and Ralph Matton was mad enough to go to Ottawa to complain. It didn't do any good. He even had to pay four extra Canadian line handler's transportation back to Port Coburn."

Q. "You mentioned that the lakes were too rough for you to tow alongside. Were you ever in some of the really bad storms we read about?"

A. "I was in one I'll never forget. We were in White City, Michigan, to load caustic soda for Cleveland. There was a storm warning. We waited for the morning for further reports. They said the winds were from the west at 30 miles per hour. I figured it's only 60 miles across the lake and though no matter how hard it blows, I'll be on the Wisconsin shore in five or six hours. We were out there for about an hour, and it started to blow, build up, and get cold, and then we got the warnings from Ft. Washington of 50 mile per hour winds. The wind kept increasing so that, by noon, we were all iced up and the radio antenna was carried away. The radar was one solid mass of ice. We had our winter metal shields on the windows so all we had were the small portholes to see out of. It wasn't long before they were covered solid with ice, and we were blind flying to the west. That wind blew all night. We didn't know if a lake ship would run us down or if they would run in between and cut the barge loose. We couldn't see a thing, but we had to keep heading into the storm. The following afternoon about three o'clock, we realized that the wind was dying off. We managed to get a hatch open and were surprised to find we were only about five miles from shore. We pulled into Milwaukee and were there for 24 hours to knock the ice off the boat. We bought wooden pick handles for the job. With that much ice, the boat would go down in the waves and not come back up. That's how bad it was. Matton had the Coast Guard out looking for us. They spotted a blob of lights towing a barge, but there was no way to tell it was us."

At the end of this story, Jack volunteered a statement that most men would not make to a stranger.

"Going up the Great Lakes on a stormy night," he said, "I'd be alone in the wheelhouse with the tug rolling and bouncing, and I'd be thinking to myself, 'What the hell am I doing here at the wheel of this tug. A little guy like me, you're supposed to be 6'6" with shoulders the same width. You know, a big raw-boned seaman.'"

Q. "I'd like to skip now from the Great Lakes to off season in New York and talk about competition for work. Could you give me a couple of examples?"

A. "In the wintertime when the canal was closed, if Ralph Matton didn't have a contract, we would have to go and pick our own work. In the morning around 7:00 a.m., we would call all the companies. 'This is the EDNA MATTON here. Do you have any work for us? We're at such and such a pier.'

"Well, if they had some extra work, they would give it to us, or they would ask us to call back later. This one afternoon, Bouchard called me about three o'clock and said, 'What's your schedule for tomorrow?' I said that there was nothing right now. He said that he had a barge to go to Port Jefferson and for us to be ready at five o'clock in the morning. That was fine. That was a nice job that would take us 12 hours to go out there, lay there for another 12, and 12 hours to get back. I got a call about five o'clock that night from Turecamo. Jack asked me if he could have us at seven in the morning. They had no intention of using me. They just wanted to know what I was doing, because we were in competition. I said that I was sorry, but we had a job. He wanted to know what we had, and I told him we were going out of town. He said, 'Oh, you got that Bouchard job.' I admitted that we did. Half an hour later, we got a call from Bouchard. 'Jack, that job to Port Jefferson fell through.' In the meantime, we got no calls from Turecamo. So, as we were crossing the harbor, who's coming toward us but a Turecamo tug with the Bouchard barge on their way to Port Jefferson. When I talked to Ralph Matton, he told me not to talk to Turecamo because he would steal a job from his own mother. That's the way competition was.

"Let me give another example. We'd be doing a small job that would take an hour or two. Some other company would call us up and ask what we were doing. I'd say we were busy, then they would want us to call them back. If for example, I were on a job for Turecamo, I would call them and tell them I would be doing a job for McAllister after I finished with their job. 'Oh, no, we have a follow-up job for you. You can't do that.' I'd get to the dock a few minutes later and I'd ask what they had. 'Well, that's all we got.' We'd lost the other job.

"Turecamo was a tough outfit. They wanted our boats because our boats had the power. I've seen us have a job, say tonight, and some company would want us to take an oil barge somewhere at seven o'clock in the morning. Turecamo would call and say they needed us for a ship job, dock or undock. We would say that we couldn't do it. 'Why not?" they'd say. So we would tell them we had to move a barge for someone else. 'Forget that job.' We can't forget it. They would even go so far as to say that they would put one of their tugs on it while we handled their job. Sometimes we would call Matton and it would take a dozen calls to straighten that out. We would ask if they had anything else for us. 'Hell no, we haven't even got enough for our tugs. Take that thing up to Albany and tie it up.' "

"We'd get to know all the offices and call them up. They would give us the work because they knew we were eager beavers. But, the other guys would just sit at the dock and wait for work to come to them. We'd bring the crew down on Monday morning to Grand Central Station. We'd head right to their offices and they'd say, 'Here, have a drink.' And if your tug wasn't in there all ready for work, they'd call up one of their tugs. 'Come over to Pier 1 and pick up the crew of the EDNA (MATTON).' They'd put you on your boat so you wouldn't have to hang around. I'm talking about McAllister now.

"Let me tell you something about Ralph and Margaret Matton and competition. We'd be laying in New York with nothing to do on a Friday afternoon. No work and they'd send you home for the weekend. Mrs. Matton would tell us that Bouchard had a job for Philadelphia that would take care of us for the weekend, but they weren't offering enough money. So instead, they paid our car fare home and we'd be back on Monday. They were very independent. They wouldn't cut prices, and I respected them for that."

Q. "Did Ralph Matton ever take the wheel of a tug?"

A. "Yeah, twice, We were taking a boat out for a trial run. It was the first tug Matton had built for the government of Peru, and it was loaded with Admirals, US officials, and guests. They said they'd like to see the boat put through a figure 8 at full speed. I said you can't do it because it's too shallow, and the suction of a tug under full power will pull right down. Ralph said, 'Let me take it.' He took the wheel and made a hard right to make the first part of the eight, but it didn't make it. She started climbing over the bottom, but we finally got it off. He just walked off, disgusted with himself. We went back to the yard. I went up to the office where Mrs. Matton had a face about four feet long. She said, 'Well, you're a nice one to take a new boat out on a trial run and put it aground.' Mrs. Matton, I didn't put that boat aground. She said, 'Oh! My God, it must have been the bullhead.'

"The next time Ralph took the wheel, the results were much more serious. It was in the winter and some of his boats were tied up in the yard. He took the wheel of one of the old steam tugs, headed for the dock, miscalculated, and ran into the middle of the outside boat. It was the

FANNIE BAKER, and he sunk it right at the dock."

Q. "As long as we are talking about speed, what is the difference in speed today as opposed to the thirties?"

A. "We used to run to New York from Albany and take 48 hours. Today we do the same thing in 10 to 12 hours."

Q. "Can you give me an example of any difference in operating then and now?"

A. "Back then no one seemed to be in a hurry. We'd leave New York for Buffalo, and they would say call me when you get there. Today, they want to know where you are all the time. This all started when the union entered the picture."

Most of our conversation took place in the wheelhouse as Jack was waiting to undock a banana boat, the RONDI. Walt Snider, the deckhand, was spreading a canvas over the bow fender of the tug. This is done so they don't leave any marks on the hull as Jack pushes against it. All banana boats are painted white, and they don't appreciate a careless tug Captain marking up their pristine hulls.

I received the same answer from Jack that I got from most boatmen when I asked the pros and cons of the work. He liked the money and the travel, but didn't like being away from home. This leads to my last question.

Q. "How did it affect family life?"

A. "I figured it was worse for the women at home. The men have the company of each other. It's lonely with children and worse without them. She doesn't have too many people to spend time with. She's like a fifth wheel. Married couples don't want her, because she's a spare tire, and they might even have to pay her way. And she can't go out with single girls because they're looking for boyfriends. They were alone when the kids got sick. They had to defend and protect them without a man to help. We missed things like graduations and birthdays. Our wives became more self-reliant than the women that had men at home each night."

In addition to helping so much with this book, Jack and his lovely wife, Dorothy, became my very good friends whom I visit as often as possible. I love them both.

Capt. Jack Maloney surfboarding behind tug, circa 1930.

P.C. RONAN, Jack's first tug.

CHAPTER 8

BEN COWLES OF BUFFALO

Very few people have been successfully involved in as many facets of the marine industry as was Ben Cowles. Born in 1863 to a shipbuilding family, while still in his twenties he became a "double-ender." A double-ender is someone who holds a license for both the wheelhouse and the engine room, which is not an easy task to accomplish.

He left his father's business to pilot ferry boats from New York City to New Jersey. Following this, he captained tugs on the Hudson River for fifteen years. He returned to Buffalo in 1901 and in the following year was appointed by his brother-in-law (the Mayor) to the position of Harbor Master. At the end of the Mayor's term in 1905, he left and founded the Cowles Shipyard Company at the foot of Genesee Street.

The firm built steel tugs and repaired tugs and barges. Some time later, he moved his operation to St. Clair Street and finally to Ohio Street. Ben also contracted for harbor, lake towing, and dredging. In addition, he towed gravel out of Palmyra, New York in 1910 and 1911 on the old Erie Canal, a fact which may have influenced him to get into the sand and gravel business. This enterprise was later sold to the Buffalo Sand and Gravel Company. In this same period, he became a business associate of Murphy and Hoffman, canal brokers.

Ben had a rather wry sense of humor, as witnessed by the following exchange between Ben and a customer with a boat on Ben's dry dock. When he looked at the bill he said that Riley never charged that much for the same work. A slight pause and Ben very quietly remarked, "Riley's out of business, isn't he?"

While doing my research, I would ask old-timers what they could remember about Ben Cowles. Invariably, they would talk about the old Lake Erie fish tugs that he converted to work on the canal. The remarks were never very complimentary. True, these boats were not exactly Cadillac models, but Ben had an answer for his critics. He liked to say, "If I can make money with a $2,000 tug, why pay more?"

These steam tugs had been used for gill netting on Lake Erie. Ben raised the after section of the deckhouse to make room for a galley. The crew slept below decks at the forward end of the tug. The Captain and mate slept in the raised section in back of the wheelhouse.

Cowles tugs were leased to the state during construction of the new canal.

It was necessary for the state to provide towing for animal drawn boats where the towpaths had been torn out. They would not be replaced. It signaled the end of horse and mule drawn boats on the canal. An interesting sidelight turned up concerning the lease of Ben's tug the WILLIAM G. FOX. The state inspectors rejected this tug as inadequate for towing, yet the next document I read showed this tug on the list of boats leased for 1917. I wonder how he managed that.

During World War I, Ben received national attention by cutting the steamer NORTHLAND in half and floating her through the Welland Canal. She was joined again at Quebec and placed in convoy duty.

There is some speculation that Captain Cowles' friend, T.V. Connors, one time chairman of the U.S. Shipping Board, suggested that he convert those fish tugs for lease to the government when the U.S. took over control of the canal during World War I. In any event, he did lease them to the government.

Fish tugs were not the only boats to be converted by Ben. In 1913, he built a tug for the Wickwire Steel Company of Buffalo. The company asked him to add benches and canopies so that they could use it for a ferry until the trolley lines reached their plant. In 1917, Ben bought the tug back, made some external changes, and put it to work as a real tug. He changed the name to the LIBERTY and kept her in service until November 1934.

For years, the only private drydock in the canal system was the one built and operated by Matton. That monopoly ended in 1925 when Matton built and floated a drydock to Buffalo and became a partner of Ben Cowles. The partnership only lasted one year, when Matton dropped out and Edwin J. Lenahan (Buffalo businessman) became Ben's new partner. This association continued until Ben's death, after which his widow and Edwin Lenahan kept the business going until 1934.

During the 1920s, Cowles/Lenahan hauled wheat in their own barges and also towed petroleum products for the Standard Oil Company. Their fleet reached its peak at this time with 16 tugs, 11 barges, and 3 lighters. Captain Cowles died alone in the Hotel Alpine in New York City on November 26, 1930. His death was discovered when an old friend, T.V. Connors, telephoned Ben's room but received no answer. Hotel employees were notified, entered the room and found the body. Sadly, there was no second generation to carry on his name.

Ben did not live to see the tug named for him launched. Perhaps it's just as well he didn't. The launching was not what you could call a complete success. The tug was literally shoved off the dock wall. The procedure was a bit unusual maybe, but it had been done before. They had to use this method because there wasn't any place at water level to launch in the normal manner. Unfortunately, this time it stoved in some plates and ribs. With a large crowd watching, it must have been embarrassing for the company.

Ben Cowles died a respected man and a rich one, by the standards of the day. He died a year after the appalling crash of 1929. He left a net estate of $246,474.45. This was an impressive sum when you consider that one winter in his early years, he and his wife lived on his own passenger boat in order to save money. When things really got rough, they had to sell some of the brass fittings off the boat to buy food.

It's sad to think there is no physical evidence that Captain Cowles was ever here. Even his shipyard location has been filled in to make a playground for the children of Buffalo. He is not entirely forgotten, however; he lives in the memory of Richard Garrity, the author of the book *Canal Boatman*, of Tonawanda, a long time employee and the person who helped me with this and other portions of my book.

Example of a Lake Erie fish tug owned by Ben Cowles, circa 1920.

Launching of the BEN COWLES by pushing it off the seawall, Buffalo, 1930.

CHAPTER 9

THE COYNES OF SYRACUSE

Technically speaking, she was a servant, but in reality she was a member of the family. She was the very last of her kind and much loved by all. Time had taken its toll. She was tired, weak and slow, but you would be too after 50 years of heavy labor. A servant yes, but never subservient and one who always carried herself proudly. Even when times were bad and money was in short supply, the family managed to dress her in the latest style, sometimes to the envy of her contemporaries.

After her retirement, the family kept her with them, but the attention she needed and the cost of keeping her healthy was a severe drain on their pocketbook. They searched for a permanent home for her, but failed. Many were sympathetic, but no one offered to help. With guilt overwhelming them and a heavy heart, they had only one choice left. Cruel as it may sound, they simply abandoned her to fend for herself. With no support, her condition deteriorated rapidly. In 1973, her end came at the hands of an indifferent government agency. She was literally torn apart by a steel jawed monster called a clam shell crane. For you see, she was the last wood tugboat on the New York State Barge Canal and was called the COYNE SISTERS.

I hope this introduction will give you some insight into the character of the Coyne family of Syracuse, New York. I doubt if you could find a family with more members actively involved in the towing business. Father, the captain, Mother the cook, one son in the pilot house, and one in the engine room. In addition, a daughter capable of standing an engine room watch, plus cousins, nephews and in-laws. This held true from the early 1920s until they quit the business in 1972.

Both sides of the family worked on the old Erie Canal. Frank drove mules as a boy of ten. Teresa worked for her uncle, Tom Hadcock, as a cook on the "Hoodledasher" [a self-propelled, steam powered, cargo carrying canal boat] ESTHER. Romance blossomed when young Frank Coyne went to work for Teresa's uncle Tom. They were married in 1903. On December 4, 1905, Isabelle was born. She was the first of five children. Tom was born on November 13, 1907 followed by Helena on July 3, 1911, Joe on November 21, 1913, and lastly Evelyn on March 27, 1917.

Before owning his own boats, Frank perfected his skills with several companies and finally became captain of the motor tanker ROCHESTER SOCONY. Even though this was a good job with a large and reputable firm, Frank's younger brother, Ross, kept urging him to get a boat of his own. He would never get anywhere if he didn't, is the way Ross put it. Ross had bought his first tug, the RESCUE, in 1921 and his second, the LOUIS BANKS the following year.

After much discussion, Frank and Teresa decided to take the plunge. They bought the tug ARIEL in 1923. This had to be a very scary and courageous thing for them to do. Frank had only a fourth grade education and neither had a whisper of knowledge about operating a business. They mortgaged their home for $500 and that was a devil of a lot of money in the early 1920s. To trade the security of a job with Standard Oil for the uncertainty of entering a highly competitive business was probably the most awesome decision Frank and Teresa ever made in their lives. They hauled grain for Hedger from Buffalo to New York for three years with that tug. In 1928, the boiler couldn't pass inspection so she was junked in Syracuse harbor.

At this time the Tice Company of New York raised the tug AUTOMATIC from Gravesend Bay, took her to Jacobson and Peterson's yard in Oyster Bay to be rebuilt. The work was about half done when Frank bought her. A new deckhouse was put on and a new engine installed to complete the overhaul. Frank named the tug after his oldest son, Thomas R.

This also seemed the right time to incorporate as the Oswego River Towing Company. Frank, Teresa and their two sons became the officers. Now the Coynes were in the real world of business. They might never match the giants such as Moran or McAllister, but they were on their way to the American dream. Daughter Isabelle told me how they loved to hold family business meetings, of course she was excluded, but she understood and accepted that. Yet the part she played contributed very much to their success.

As soon as they bought the ARIEL, her mother went on the boat to cook. This meant someone had to take care of the younger children when they were in school and when the boat was working. From the time Isabelle was 17, she filled this roll even after she married Paul Sherman, who also worked for the Coynes. This kind of devotion to a family's needs was to me a beautiful act of love and caring. Starting the summer of 1923, all the kids except Isabelle spent their summers on the tug. What an adventure that was for them. They were the envy of all their friends. With such a small boat, only 80 feet long and a crew of nine, I was curious as to where the kids slept. On the ARIEL, they hung hammocks in the galley. On the THOMAS R. they were squeezed into rooms. With quarters this close, it's difficult for adults to get along. It must have been chaotic at times with four noisy active kids, ages 6 to 15. Needless to say, the crews had to be very tolerant. They did have chores to do, such as washing dishes and polishing brass. Sometimes to get them out from under foot, they would be sent to the bow rope locker to paint. Mama and Papa never did go down to check the work. Of course there were times when a little discipline was called for. Frank was quick tempered, but over it quickly. He did the yelling while Mama did the hitting. She had one rule: they were not allowed to cry. Everything considered, it was the kind of togetherness that is seldom matched today. It was even closer than a farm family, because the farthest apart you could get was less than 80 feet.

In the beginning there were rough years. Frank would have to borrow money each spring to get started for the season. What success they had was due to hard work on the part of both Frank and Teresa. Again I quote daughter Isabelle, "The two of them, they were great workers. Mom was a wonderful cook. And of course, everyone gained 35 pounds every summer. She baked pies and cakes and cookies. Everyone loved her cooking and she saved in ways that big companies couldn't. They put their noses to the grindstone and that was it. They slaved and slaved and slaved. This was their life." In those days, there were no two weeks on and then two weeks off as it is today. Even after the union came in, they had to work the same way just to survive. It always bothered Frank that Teresa had to cook in order for them to make it. Mother Teresa saved money by washing the crew's linen, in addition to all the laundry for the family. Large companies send their linen out to a commercial laundry.

About the time they had the THOMAS R. ready to go, they got a contract from Bouchard to haul oil to the steel mills in Lackawanna, New York. This cues me into a story about Frank, which the family loves to repeat. A barge caught fire while Frank was at the steel mill docks. Not waiting for the local fire department, he and his crew put the fire out at some risk to themselves and their tug. This impressed the steel company so that they insisted the Coyne haul all the oil to them. The prime contractor was Bouchard, but Coyne was hired to provide the tugs. They ran oil to the mills until 1942, at which time Bouchard lost the contract.

I'll stop here and relate a couple more stories about Frank. These were told to me by Paul Sherman, son-in-law and captain for the Coynes. The first one happened on the ARIEL. In the days of steam, boilers frequently sprung leaks that sent water into the fire box. When this occurred, the fire had to be pulled, the boiler cooled down and then someone had to crawl in the fire box to caulk the leak. This isn't too difficult a chore if the box is completely cooled down. When you own your own tug and operate on a shoestring, you can't afford the luxury of waiting too long. While still quite warm, Frank would throw some burlap bags in on the grates and crawl in. Paul said he wouldn't have gone in there for a million bucks. Besides, the job normally belongs to the fireman or engineer.

Another measure of the man can be seen in an incident that happened in Waterford, New York. Frank never spent much money on himself, but he had just bought a new pair of shoes for which he was rather proud. On the approach wall into Lock 2, a small boy was walking along watching the boats when he slipped and fell in. Frank stopped the tow and with no thought to his new shoes, dove in and saved the boy. That boy years later worked for Frank's son Joe as a mate. Unless you have lived through a depression or have known hard times, it's difficult for me to convey to you the significance of being able to buy something as mundane as a new pair of shoes. I often wondered if Frank learned to swim in the same manner he taught his children. He tied a rope around their waist and heaved them over the side.

Talking with the family, I discovered that Frank always worried about not having any education or polish. He was never comfortable around educated people. His wife had taught him how to read and write, but he didn't need to worry. I was told that he held his own very well with bankers and other business men. He won the respect of all he dealt with. I've noticed throughout my research that many of the boat people overcome the lack of formal education with something I'll call native know how. If Frank and his family had one fault it was putting too much trust in some of the folks they did business with. Even if it was in the days of a man's handshake as his bond, caution was still the watch word. Unfortunately, this philosophy contributed to the ultimate failure of the company, when son Joe tried to match wits with Sears Oil of Rome, New York. More on that later.

Back on the THOMAS R. they ran her alone until the purchase of the tug I opened this chapter with. The COYNE SISTERS started out as a Navy tug. It was 1935 and she joined the THOMAS R. in hauling oil to Lackawanna until the end of that contract. After this they went to work for Sears Oil. Sears is a small oil distributor operating out of Rome, New York. The mildest comment I ever heard was that he was a shrewd operator. In any event, the Coynes were not in the same league. Two years after the purchase of the COYNE SISTERS, they made their only attempt to enlarge the company by buying the third tug. This one was named after his youngest son, Joseph, and had to be sold two years later for lack of work.

They had no work during 1946-47, so it seemed very strange to me that Frank at age 65 decided to try to make what would have been the biggest deal of his life. With the war over, many surplus tugs were put up for sale. With no real prospect of work, he attempted to buy an Army tug for $85,000. This was an enormous sum in 1946. I can only guess at his reasons. One may have been so he could work in the winter because it was a steel tug. Another reason may have been his desire to help his sons get a foundation with a modern tug, rather than nurse a few more years out of the ancient wood tugs they owned. It wound up as an academic exercise anyway. The government insisted that Frank surrender all his assets for completion of the deal. Frank decided the risk was too great and backed out. To see how it was done, not once but twice, they should have talked with Marty Kehoe who had no money at the time he bought two surplus tugs.

The next ten years were bad ones for the Coynes. They got a few jobs such as shifting barges for a construction company laying pipe lines in Lake Ontario. During this time, son Tom worked for the State of New York on a dipper dredge while his brother Joe started selling real estate. Joe became quite successful at it. Tom also worked for Kehoe, McAllister, and Bushey, but couldn't hold a job very long with any of them.

During short periods of slack work, Frank would pay family members of his crew a token amount of money to at least put some food on the table. There was no unemployment insurance in those days. He always tried to be fair. He wanted to keep what he called his key men on the payroll. That would be Paul Sherman, Rube Coyne (nephew), and Joe McHugh (son-in-law). His sons, as officers of the company, turned thumbs down on that idea, so everyone had to scramble for work. They were forced to sell the THOMAS R. in 1953.

In spite of the set backs, the pull of the towing business was still very strong. They were back in business in 1955. It was a tragic year for the family. Their beloved father died and they went back to work for Sears Oil. You may feel that tragic is a strange word to use in conjunction with a contract with Sears Oil. I could have used calamity, catastrophic, or disastrous. At three different times the Coynes worked for Sears. After each period, they felt they had been ill used. Other terms were mentioned, but I hesitate to use such language. The wonder, to me, is that they kept going back to them.

At first, the yearly contracts were satisfactory, but in time they had a subtle way of changing, more in favor of Mr. Sears. Let's look at the last phase of the Oswego Towing Company. By 1958, they were once again out of the towing business. Joe went full time into real estate. His brother and the men found jobs where they could. The COYNE SISTERS was kept tied up in Syracuse harbor. The Coyne story would have ended here, except for the entrance of a pretty 21-year old redhead named Shirley Stills. In 1962 she went to work for Joe and by August of that year she became Mrs. Coyne. This of course put her in close contact with Joe's family and all the boating stories. Shirley was especially taken by Joe's mother and listened enthralled by what Teresa had to say. She was fascinated by and much admired Joe's mother. Unfortunately, she was only hearing the lighter side and none of the dark. Time, I have found, has the tendency to dull the harsh realities of life. She thought it would be great to have the COYNE SISTERS run again. Joe was not anxious to start up the towing business again. From time to time, the phone would ring and someone would ask Joe to take a small job. He always turned them down. She kept after Joe and suggested they combine a vacation with taking one of those small jobs.

Her timing was perfect. Real estate had gone into a slump. Another call had come in to tow five small wood barges from New York to Weedsport. Joe caved in, rounded up a crew, fired up the tug and headed for New York. The trip down was all that Shirley dreamed it would be. Shirley said, "The price we got for that little job turned me on." Decision time again. They might be able to get more work. If not, they should try to sell the COYNE SISTERS because it was becoming a financial drag. They talked to a broker, and of course he wanted to sell them a fleet. He did have two small tugs that would be perfect for them, so the sales pitch went. They were Connors boats, the HARRY R. and the ANNA L. The asking price was $62,000 for the former and $50,000 for the latter.

The fever caught hold and it wasn't long before the predictable call to Mr. Sears. They of course could not afford to buy them without help. You can guess the results of the call. Sears thought it was a grand idea to buy the tugs and have Joe work for him again. He'd been after Joe for a long time to get back in business. You can almost see him gleefully rubbing his hands together, a slight smile on his face and he thinks – Wow! I've got him again.

You just know that Joe is going to come up, "four dollars short and two hours late," once more. Sears bought the boats and allowed them to be named FRANK COYNE and TERESA COYNE, and took the COYNE SISTERS as collateral.

Joe and Shirley figured they would get the best of the deal this time

Capt. Frank Coyne.

126

Theresa Coyne.

because they had a plan. They would work for Sears for two or three years and then negotiate a better contract with someone else. So much for dreams. Once again Joe had underestimated Mr. Sears. The lesson the Coynes never learned is a very old one. Nice, honest, hard working, trusting people always come out second best when dealing with shrewd folks like Mr. Sears.

It's interesting that Shirley, without a trace of bitterness told me that everyone blamed her for the failure because of her curiosity and enthusiasm. I don't hold with that. Given time and with no Shirley, I feel Joe would have gone back to the business with the same results. It's difficult for an old sailor to "swallow the anchor."

I asked Shirley how the contract worked. They were paid by the trip and Sears would deduct the mortgage payments. It became evident, very quickly, that this was not the way to operate because they didn't receive a receipt for the mortgage payments. From that point on, the payments were handled separately. Let me quote Shirley again, "You had to be right on your toes every minute. He and I used to get into it. Oh boy! Because he never wanted to pay. And so every year, of course, with the union contracts and everything, we'd have to negotiate. Sit there, and he was a great one to wear you down. So we'd sit there, sometimes till two o'clock in the morning, trying to agree on a rate." My next question was, "Did you ever come out on the losing end?" "Oh, yeah, that's why we ended up abandoning it because in the end it would cost us money to work for him."

One thing might have saved them. If they could have worked the tugs in the winter in New York. That was almost impossible for two reasons. The horsepower of the engines was too small and they didn't have the necessary contacts in New York City. They were really working part time, because the canal was closed all winter.

Shirley did say that they should have had a top flight marine lawyer. They used a lawyer that Frank's Dad has used over the years. Loyalty on Joe's part may have been the cause of some of their grief. Shirley felt he was half asleep and didn't ask enough questions during the negotiations. To be fair to the lawyer, the pressure of time may have had a bearing on his opinion of the contract, even if he wasn't convinced he had made the best deal possible for Joe and Shirley.

At times Sears would renege on a payment and there would be a running fight to get him to pay up. He always kept them off balance. He had an ace up his sleeve that they knew about, but he had promised he wouldn't use. God! How could they have been so naive. They were making payments on the FRANK COYNE only. They never did own the TERESA COYNE.

Although he said he wouldn't, any time work got slack, he would make them use the TERESA. This meant they had to pay the charter rate while the FRANK sat idle. In today's vernacular, this is called a "GOTCHA."

At about this far into my interview with Shirley, I almost began to doubt what I was hearing. I could not imagine anybody, not even a greenhorn, which Joe wasn't, allowing themselves to get into such a position. They were entirely at the mercy of Mr. Sears.

The plot thickens as the villain strikes again. In 1970, they didn't have a contract because in Shirley's words, they were getting only 50 dollars a day profit and it wasn't enough. So they went to work for someone else, which of course, upset Mr. Sears a wee bit. Mr. Sears pulled another ace out of his sleeve and brought the United States Marshal down on them. He filed a complaint that they had defaulted on their mortgage payment. So the Marshal padlocked the boat and slapped a plaster [slang for an official seal to prevent use of boat] on it in Waterford. They had made the payments and had the receipts to prove it, but they would have to go to court to get it settled. Mr. Sears knowing their circumstances, bet they would not take that step. Again, Mr. Sears had won. They hired a lawyer, haggled for three days and wound up working for him once more.

I asked Shirley if it wouldn't have been worth the effort to go to court. Her answer and I quote, "He [Joe] didn't like to fight, he believed if it could be resolved without going to court, then why go to court. And I think he was a little bit afraid of this wishy-washy agreement we went into, that was so one sided. How the court would interpret it, nobody could know. It was my contention that the big guy just doesn't squash the little guy and get away with it. I mean, when it's flagrant. I still think that any court would have looked on this case in our favor, but Joe didn't."

I'll relate one more Sears story and then go on in a lighter vein. The exact time of this exercise is a little fuzzy. It varies with whom you are talking with. About 1967 to 1969 four towing companies started to double tow oil barges. They were, Bushey, Moran, Kehoe and Coyne. Each one will claim they were first, but that's not really important to the story.

I don't think we can fault the companies for trying to find ways to increase their profits. In theory, they could move twice as much oil for about the same costs. There were some up-front costs, such as fitting the lead barge with a power unit to help steer the longer tow around bends, plus the unit was used to pull the lead barge out of the locks. These units are called bow thrusters. One way to describe them would be to compare them to a giant outboard motor. Most were controlled from the tug pilothouse.

There were two reasons why this idea was not successful. The most important one was the fact that, with rare exceptions, terminals could fill or pump out only one barge at a time. Result: No saving of time for this part of the operation. Secondly, because of double locking, travel time was longer. The Mohawk River part of the run was not much of a problem as far as traffic and room to maneuver, except during high water when the faster current could raise hell with the tow. The rest of the canal was something entirely different. Narrow channels, bends, and small boats really tested the skills of the men in the pilothouses. Even under the best of circumstances, I've seen amateur boatmen give tug crews fits. This was probably the only time when Joe Coyne stood up to Mr. Sears and said, "Stuff it." One final note. Joe had one more cross to bear with this fiasco. His bow thruster could not be controlled from the bridge. He had to signal someone on the bow of the barge And that is one lousy way to run a railroad. Sears, no doubt, saved a few pennies on the installation.

Shirley didn't come up against the real world of towing until they made their first run with the FRANK COYNE. The trip on the COYNE SISTERS was the honeymoon, now tugboat life began in earnest, with a near mutiny.

I'll use Shirley's own words to tell the story of the "Great Pancake Mutiny." "I was very new to this thing. I never had to deal with six or seven men. Number one, they didn't like the idea of having a girl on the boat in the first place, let alone a young girl, let alone the Captain's wife. We were in the process of signing all these papers in Albany for the boats, pick up grub, so we were running late. By the time we got back to the boat, it was 5:00 p.m. and the normal time to eat. We've got to take off, so Joe says, why not fix pancakes and sausages for supper – that would be good and you can do it fast and they'll understand because they know we're buying the boats and running around." This may have been the dumbest words Joe ever spoke in his life. Back to Shirley, "So I did. I didn't see anything wrong with that. So I made pancakes and sausages – Oh dear! Oh dear! – the guys walked in and they looked around and sat down. The two Swedes looked at each other and the roof came tumbling down."

"There was no way – they knew they shouldn't have a woman on this boat – they knew it would never work out – if this was the way things were going to be, they were going to get off. Oh, it was bad! I was heart broken. I didn't know, we'd been with this group all week, this was all new to me. I had never cooked before – just on that little trip and that was all people we knew."

Left to right: Tom and Joe Coyne.

Mate Cliff Arnold.

A little panicky by then, she headed for the wheelhouse to Joe. Refusing to cry before the men, the dam burst when she reached Joe. He came down to soothe everyone's feathers and they finally ate. Shirley hadn't been told by anyone the strong traditions on a tug. Cardinal rule #1, you do not serve breakfast type food for supper, period.

To apologize for her faux pas, she made a special midnight desert. Being a good cook, saved her from having any more confrontations with the crew. She adapted to the extent of preparing special food for a Jewish lad in the crew.

I asked her if she had any other clashes with the men in other areas. Only one that was big enough to stick in her mind. He had been a captain, but was working for them as a mate. Talking with other boatmen, they were not surprised at what he did. As Shirley phrased it, they had little "pleasures" in the galley such as real glasses. They were not to be taken on deck. No one did except the mate. After the first batch disappeared, she bought 12 more in a holder that sat on the back of the table and carried a renewed warning about taking them on deck. Of course, the mate ignored the rule. By the time four were missing, the crew began telling the mate he was headed for trouble. In a joking manner, Shirley said that he'd better find those glasses. The mate turned, stomped into the galley, grabbed the remaining glasses and heaved them overboard. To add insult to injury, he threw over a sugar bowl and two little condiment things that Shirley was partial to. At this point she told him he was being very childish. And that is where he made his fatal mistake. He told her, if she didn't like it she could get off the boat. Oh my! That was not the appropriate thing to say to the owner's wife. Result – Bye, bye matey.

One more incident tied to the galley is worth repeating. On one trip "outside" (on the ocean) the weather turned sour and they were bouncing around pretty good as Shirley was making a favorite dish – clam chowder. At the precise moment she was putting in the pepper, the tug gave a lurch and in goes the whole can. At that point she was not about to start a new batch. so she fished out the can and as much of the pepper as she could. Holding her breath as the crew sat down, she watched them start in . After the first couple of spoonfuls, the engineer looked at mate who looked at the deckhand and almost in unison they said, "Jeez, Shirley, this is the best clam chowder you've ever made." And that is how legends are born. After the truth was told, it became an inside joke. Shirley did make one major change in the galley. The "black monster" (cook range) was too hot for her in the summer, so they installed a bottled gas cook top. I know some cooks who would like to follow Shirley's example if only the owners would part with the money to buy one.

An area where Shirley did take a lot of kidding was in her choice of color for the boats, turquoise. It's not exactly the color befitting the rugged image of tugboating. The lock tenders were always asking when she was going to put up the curtains.

All the boatmen I talked to agreed she was a good sport and always turned her hand to help. She painted, steered, and sometimes handled lines.

This looks like a good place to stop and introduce some of the people that helped me and their relationship to the Coynes. The first is Captain Paul Sherman who, as I wrote of earlier, was married to Frank Coyne's oldest daughter, Isabelle. Paul started his career as a deckhand for Frank on their first tug, the ARIEL. Except for a few years with Bushey, he spent his entire life working for the Coynes. Paul liked working for Bushey because they worked year round. North in the summer and south in the winter. He might have stayed with Bushey until retirement except a new boss pushed a little too hard one day. He tried to pressure Paul into running in high water when he didn't think it was safe. I'll quote Paul, "We were going to Sacket's Harbor on the eastern end of Lake Ontario. It was the first load in the spring, and I went into the lock at Phoenix with six feet of water running over the dam with every gate open. I was wide open going into that lock to keep from going sideways because of the fast water. The lock tender said that in twenty years I've never seen a tow come in here with that much water running, and I said you won't see it again, I quit!" Bushey's man said go, Paul said no.

The other side of the information coin (excuse the pun) was Rueben Coyne (Frank's nephew). A delightful person to talk to, but he had very little to tell me. I believe people go through life in one of two ways. Some observe and absorb all the sights and sounds that life presents to them. The rest zip through with tunnel vision and plugged ears. Rueben was the latter type. I sincerely think he wanted to help but was unable to. Two other long-time employees were Cliff Arnold of Central Square, New York, and Jessie Wimett of Port Byron, New York. For a time, Cliff was related to the Coynes in that his sister was Joe Coyne's first wife. Both of these men still work for Sears Oil. I thought relations would have been strained between Cliff and Joe, after his sister's divorce but Cliff said no. Although born in Syracuse, Cliff spent part of his youth in Kansas City where he had a chance encounter with a fortune teller. She told him he would work on the water, not any water but shallow water such as a canal or river. At the same time she told his brother that he would be a farmer and than a contractor. Both boys didn't put faith in her predictions. But, it all came true.

In 1937, at the age of 19 and in the middle of the "Great Depression," Cliff had no job or the prospect of one. Through his sister he was given a job as a deckhand. His next statement caused my eyebrow to raise. Cliff said Mrs. Coyne (Teresa) had some reservations in that he might not be able to "cut the mustard." Was she joking? Cliff was big enough to lift a tug clean out of the water. She, of course, was wrong. He also worked as an engineer. I have met a few men like Cliff who are called double enders. They can work in the wheelhouse or the engine room. And like the others, there comes a time when a decision has to be made as to which place to settle down in. Everyone I met finally opted for the wheelhouse. Usual reasons – oil, grease, and noise of the diesels got to them.

Cliff is winding up his career as the mate on the Sears tug MID-STATE 2 (formerly the FRANK COYNE). Tugs are named in many ways, some a little strange as you will see in other chapters, but MID-STATE 2 is a logical, no nonsense name as you might imagine Mr. Sears would choose. His business is in Rome, New York – the middle of the state.

As I talked with Cliff I can't help but think he would look more natural sitting on a tractor rather than shoving an oil barge up the canal. Part of that feeling is caused by looking at all the photos he loaned me. He is wearing farmer overalls in all of them. I wish to thank Cliff for some of the stories about Joe's brother, Tom Coyne. Two brothers couldn't have been more different.

Tom was a collector of anything that floated on the water, and it all wound up in the engine room. Cliff decided the fuel tanks needed painting, but getting to them was another matter because of Tom's "treasures." Enlisting the aid of the oiler and the deckhand, they devised a way to clear the engine room. When Tom came on watch, one of them would look to see which side Tom was steering from. If he was on the port side, they would shove all the junk off the starboard side, and vice-versa. It didn't take long to complete their foul deed. It was some days later before Tom found a reason to go to the engine room. Everyone held their breath in anticipation of his reaction to the new look in the engine room. It wasn't long in coming. On the deck again, he spotted Cliff and in a very calm voice commented how neat it looked. But after a slight pause and a woe begone look on his face he said, "Cliff, it took me ten years to collect all that wood."

Another time while steering near May's Point [near Lock 25 and the

Montezuma Marsh and Wildlife Refuge], Tom ran the tow aground. It took a while but they broke free without added help from another tug. A few days later Tom had a friend aboard riding for a short distance. Tom related the grounding and the friend naturally asked how the devil he did it. Like this, he said, and without a moment's hesitation, Tom swung the wheel over and drove the barge into the mud in exactly the same spot. There you have Tom Coyne in a nut shell. He certainly marched to a different drummer. Someone once said in describing the brothers that Joe Coyne was a go-getter while Tom was a go-letter. Tom was a dreamer, deeply religious, seller of Bibles and strangely, a devotee of the occult. The last was my downfall in my interview with him and his sister Evelyn, who was also interested in the occult. I let it slip that I had done some reading on the subject. Wrong thing for me to say. I never did get the conversation back on towing. Many said that Tom's only salvation was the fact that he worked for his family. I should make something very clear. There is nothing wrong with any of these traits, as long as you can lay them aside during working hours. From what I have observed, not devoting 100% of your attention while in the wheelhouse can lead to some embarrassing if not dangerous situations. One last story and perhaps the best about Tom. An unwritten law of any type of transportation; except in an emergency or breakdown, you keep moving. Tom broke the rule. It was shortly after World War II and Tom was tired of all the pressure of pushing (again, sorry about the pun) so hard for those years, that he gave into an impulse. There was little time for small pleasures. While pushing a loaded barge, Tom stopped the tow, had the deckhand lower the life boat, and sent him to pick water lillies! You must admit that Tom Coyne did not fit mine or most folk's image of a tugboat man. It didn't fit his father's image either. He was furious and fined Tom for the time lost.

The last person to talk to me about the Coynes. He was Captain Jessie Wimett, Jr. I hope he won't mind me using junior. Neither he nor anyone else uses it, but it helps me because his dad was also a boatman and an employee of the Coynes. Jessie Jr. was the third and because he chose not to marry, the last Wimett canal boatman. A quick meeting and you might describe him as just an ordinary down to earth type of person. True, but he possessed a rare trait. He is refreshingly tolerant of other folks foibles and tries to see each person's point of view. His comments on Sears vs. Coyne shows that.

Jessie was born in Jersey City, but sure didn't stay long. From the hospital he went directly to his mother and father's barge where he lived until old enough to go to school. At that time, his parents bought a house in his mother's home town of Port Bryon, New York. It was the place his mother and father met, courted, and married. It was not unusual for boats to be frozen in during the winter, especially in the old Erie Canal. What better way to spend a long cold winter than courting a pretty girl. Jessie finished high school before following in his father's footsteps

Jessie's first job was decking on the F. Y. ROBERTSON (Dwyer tug) on which his father was the captain. He stayed on her until she burned up in 1939. Both he and his father began a long career with Cargill Grain. They stayed until the company went out of business in 1960, a casualty of the St. Lawrence Seaway. Sometimes they were on the same boat, but most times each was a captain of his own boat. Jessie did take a few years off to attend World War II.

When Cargill Grain folded, Jessie's father finished out his career on a state tug out of Syracuse. For a couple of years, Jessie worked for two companies, now defunct – Valentine Oil and Oil Transfer. In 1963, he went with the Coynes on the old COYNE SISTERS then moved over to the FRANK COYNE (now MID-STATE, 2) and is still there as captain. Jessie Sr. also worked for Frank Coyne on the ARIEL. 1963 marked another change for Jessie. He quit working in the winter and collected unemployment until the canal opened in the spring. Many canal boatmen as they get older decline to work in the winter.

At this point, I asked him about the alleged bad blood between Sears and Coynes. His answers were given slowly and matter of factly and I quote, "Sears done everything he could to make money for himself, that's only a normal thing, I suppose. Coyne always complained he wasn't making any money, that's a normal thing too." Jessie continued in a similar vein when I asked if it was difficult to get equipment out of Sears. "He [Sears] don't like to spend no money, of course, that's a characteristic of most of these outfits now. They don't take care of their boats like they used to."

Let's look at the final days of the Oswego Towing Company. The 1971 season started on a sour note. After completing their annual spring overhaul, they headed for Albany to pick up a barge. Just as they got to Albany, they had to shut down the engine. Investigation showed contaminated oil which had scored the cylinder walls. They had just put new oil in. Naturally Joe screamed at Mobil and they sent a crew to drain, flush, and refill the system. Shirley told me it cost them $4,000.00 to repair the engine and expected Mobil to pay. Of course Mobil refused, which led to the inevitable, a law suit. Search for parts for the engine forced them to lose two months that season. Mobil offered them a settlement, which they turned down. When I talked to Shirley in 1981, the case had finally gone to court. Shirley lost.

1972 was the last year they operated. Joe took sick in July and died on October 14. Shirley finished out the season at which time Sears bought the FRANK COYNE from her.

Discounting the slack years, the Coynes survived for 49 years and I purposely chose the word, survive, because for many of those years they barely made it. But the important question is, were they successful? How are we to measure that? The family had a good life. Never rich in a monetary sense, never poor in a moral sense. They were a large, closely knit family, with the usual differences and difficulties. I choose to think they were a success in a way impossible to measure by dollar standards.

I'm adding a footnote to the history of their beloved tug, the COYNE SISTERS. In 1968 she was sold to one James S. Supley for $1.00. He was going to use her for odd towing jobs. Quote Shirley, "James Supley was not as bright as we had anticipated." An understatement would be that he didn't know a hell of a lot about tugboats, especially old wood ones. He rounded up a crew, took a spin around Syracuse harbor, tied her up, and went home. Two days later, Joe got a call from the state that the COYNE SISTERS was sinking fast. Joe rushed to the harbor but she was past the stage of using the tugs own bilge pumps. With the help of some large pumps from the state, they re-floated her and got the leaks stopped. Old wood boats need lots of tender loving rare to survive. Whatever great plans Mr. Supley may have had, we shall never know. Shortly after the sinking, he sold the tug to a Mrs. Ackerman who intended to live on it with her three children! You might say Mr. Supley took advantage of Mrs. Ackerman's ignorance of what she was getting into. One fact she probably didn't know was the state rule that you can't leave a boat tied up to a state dock forever. You must move it from time to time. Most times the rule is not rigidly enforced, but in any event, she found someone to help her take it out. Bad decision. She had a repeat of what happened to Mr. Supley. That was positively the last voyage of the COYNE SISTERS, and the end of the Oswego River Towing Line.

CHAPTER 10

JOHN E. MATTON & SON, INC.

I couldn't have written this chapter without T. Emmett Collins. Emmett, a brother to Margaret Matton, entered the business as a laborer in 1929 and finished up as Vice-President until the business was sold to Bart Turecamo in 1964. More about Emmett later.

The founder, John E. Matton, was born in Schuylerville, New York, on April 2, 1879. As a boy he and his family moved to Waterford, New York. He received his early education in Schuylerville and later attended Christian Brothers Academy in Albany, New York. The senior Mr. Matton will always be referred to in this book as "John E.," for he was never called "Johnnie" or "Johnney." No matter who I talked to, they would say "John E." Although his son was Ralph E., his middle initial was never used.

When doing research such as this, there are gaps that leave us with a mystery. One of these is how and where John E. learned the trade of caulking boats, which of course led to his career of boat building. We start with John E. owning a boat yard when he was 21. He had to apprentice to someone and he would be barely out of his training and with a new bride. Suddenly he had enough money or borrowing ability to start his own business. How he did it I don't know but we will accept it as fact and go on from there. His boat yard (the first one) was three miles north of the Waterford side-cut locks and a few hundred feet above "the upper two locks" on the old Champlain Canal.

In 1897 John E. married Edna Coons of Waterford. Their only child, Ralph, was born on February 9, 1899. His wife was a real helpmate in the early days. To save the caulkers' time, she and John E. rolled the cotton and oakum in the evenings after supper. They would do enough for the next day's work.

Also in those days boats were sometimes built and repaired right next to the canal in what might be construed as a poor man's graving dock. The state allowed them to cut through the canal bank and install wood gates. Behind the gates they would dig a large trench and the dirt became the walls of a primitive graving dock. John E. spent the winters building barges and in the spring when the canal was filled with water he could float his boats out. It was a long time before they got beyond a hand-to-mouth existence. Frugal habits were formed in those days which were sometimes hard to overcome even after they became successful. John E. was

139

able to loosen up and enjoy his success in later years. His wife still retained some of her earlier frugal habits all her life.

The construction of the new enlarged Barge Canal forced John E. to find a new location. He waited to the last possible minute to make his move. The first construction on the new canal was started practically in his back yard, at Fort Edward on April 24, 1905, but John E. didn't move to his new location until 1916. He chose an excellent place, however, on the Hudson River at Cohoes, where the traffic of the entire canal system would pass by his door. He knew he needed one item to make his boatyard complete, and that was a floating drydock. He had part of the river bed blasted out and built one 44 ft. × 110 ft. From 1916 until it was torn apart in 1964, John E. had the only privately owned drydock between Cohoes and Buffalo. There was one in Buffalo for a few years, operated by Ben Cowles, and John E. was involved in that. In fact, he built it, towed it to Buffalo, and was a partner with Cowles for a year.

Matton's reputation grew over the years. Many people, including the competition, told me he built the finest wood barges money could buy. That tradition was carried on by his son Ralph, when he began to build steel tugboats. Their repair work was held in the same high esteem.

In 1917, Ralph graduated from high school and in doing so he upset the whole village of Waterford, New York. The graduation ceremony was always held in the Baptist Church, something that troubled his staunch Catholic soul. In 1917 students didn't "buck the system." Well, Ralph did. He made an issue of wanting his parish priest on the stage with him. The school agreed and, to avoid future problems decided to move the ceremony to neutral ground – the town hall.

The following year John E. hired a young lady who had just graduated from a business college in Troy and who was to have a profound effect on the Matton family. Her name was Margaret Collins, hired to be John E's secretary. In a short time she would become his office major-domo and a very important member of the firm.

1922 was a great year for John E. His son Ralph graduated from a prestigious engineering college (Rensselaer Polytechnic Institute, in Troy, N.Y.) and joined the company. It also seemed the right time to incorporate and from then on it was JOHN E. MATTON AND SON. The father was enormously proud of his son until the day he died. Ralph never failed him. John E. sold the business to the Corporation for a $25,000 bond and 1000 shares of stock. That may sound like an impressive amount of money, but

Wood barges built by John E. Matton, circa 1920.

Early Matton tug (1924). Notice square wheelhouse—easier to build.

U.S.S. LAVALLEE towing drydock to Ben Cowles, 1924.

Raising sunken barge, tug MATTON 10, August 9, 1928.

Making the movie "A Girl on a Barge," using tug P.C. RONAN and a scow for a camera platform, 1928.

Ralph E. Matton's homemade crane.

Margaret Matton.

Ralph E. Matton.

Left to right: Margaret, John E., Edna and Ralph.

Wood sub-chaser built during World War II.

Sixty-five-foot Army tug built in 1945.

151

Matton Shipyard, Cohoes, New York, 1949. Oil transfer barge on the ways.

Launching oil transfer barge #31, 1949.

Tug EDWARD MATTON with removable pilot house, storm shutters on main pilot house in lowered position, 1951.

One blow of the ax will launch another tug.

Raising Navy bomber from Lake Champlain, July 3, 1957.

General Motors' 12 cylinder engine. Compare man to engine, 1956.

Lowering engine into tug MATTON 25, 1956.

Example of a "chine" built hull.

remember it was only paper. John E's real assets at that time were exactly $585.94 and one barge in the process of being built.

In 1924 they started the towing part of their business with the purchase of the tug, VERONICA L. (renamed R.E. MATTON) from David Roberts of Champlain, New York. While checking the history of the VERONICA L, I discovered something odd. Three years *after* he bought her he formed the Steamtug R.E. MATTON Corp. with David Roberts. The strange part was the fact that John E. held no office and only one share of stock! One year later John E. bought Roberts out. Partnerships were not unusual in the days of wood barges, but normally they were used as a form of mortgage, held until the barge was paid for.

Although the VERONICA L was not old or in poor condition, she was the exception when talking about all the tugs Matton bought. Buying old boats and putting them in shape was more the norm. And why not? With a drydock and skilled workers they had a leg up on the competition.

Between 1924 and 1936 they bought 16 old tugs – 15 wood and one steel. An interesting item is the length of time they owned these tugs. They kept the WM. R. HULZISER only one year, yet the H.A. MELDRUM, an old steam tug converted to diesel, was part of the fleet for 33 years.

Information about how much boats were bought and sold for is very difficult to obtain. For example, I was naive enough to think that a Coast Guard Bill of Sale or Abstract of Title would contain the dollars paid for the vessel, but such documents don't show these data. Some say that the vessel was sold for the sum of $1.00 & O.G.&V.C., an archaic term meaning "Other Good and Valuable Considerations." Some will be listed as a Fleet Mortgage for a "zillion dollars," so you still have no idea what a particular vessel sold for.

In 1925 an incident almost ended Ralph's career before it had barely got started. He was working with a crew on a wood barge when the blow torch he was using became partially plugged. In an attempt to unclog it, while it was still burning, the top blew off drenching Ralph with gasoline. His clothes were instantly set on fire. Homer Folger rushed to his aid, tearing Ralph's clothing off. Homer's hands and arms were severely burned by this action, but he probably saved Ralph's life. So fierce were the flames which enveloped Ralph that his father thought one of the boats had caught fire. Not waiting for an ambulance, they rushed him to Leonard Hospital (across the river in N. Troy) in an open touring car, with Ralph standing up in the breeze in an attempt to ease the pain a little. It was thought for some time he would not survive. Burn treatment was rather primitive in those days. Luckily, he had not inhaled any fumes. Ralph recovered, but most of his body, except for his face and hands was scarred for life. As we shall see, the Leonard Hospital would become a focal point for the Matton family in later years.

By the middle 1920's, the business was doing well enough for John E. and Edna to start taking vacations. Miami, Florida, was their favorite place, where John E. could play golf. They also took some cruises in the Caribbean. These vacations were never more than three or four weeks long. It's not that Ralph couldn't handle the business, it was more of a case of boredom. Back home again, the normal routine prevailed. John E. stayed in the office and Ralph headed for the yard, where he felt most comfortable. John E. was the planner, while Ralph was the doer.

John E. and Ralph were much alike in appearance, but definitely not in demeanor. Both were heavy in weight and bald most of their lives. Even the oldest employees addressed John E. as Mr. Matton, while his son was always called by his first name. Ralph far outpaced his father in the "short fuse" and swearing department, and both enjoyed good food as their photographs show.

By today's standards they both maintained a unique incentive program for their employees – "Those that worked kept their jobs." Because they had a small workforce, they were faced with the problem of scheduling to keep the men busy. At different stages of building a tug, different trades are required. As an example, burners [a burner uses oxy/acetylene torch to cut steel plate] are needed in the beginning but not painters. So that they would not have to be laid off, they were cross-trained in more than one trade. The burners might become painters or pipe fitter helpers. This policy wasn't philanthropy on the part of the Matton's; it made good business sense.

John E. approached the necessity of paying bills in a way that made good business sense to him, but not to others. John E. had what some would call an admirable trait. He believed that a bill should be paid on the day he received it. The last thing he did each day before he headed for home was to drop off checks at the post office. Ralph may have admired this trait in his father but he could not embrace it. In Ralph's time the bills were too large to keep this practice going. The more common practice was and is now to pay at the last minute, while you keep your cash in an interest-bearing account.

Ralph felt more at home in the yard because all his life he was a very physical man. His nickname was "Bullhead," but never to his face. He was apt to grab a tool out of the hands of one of his men, and show him how it should be done. That action sometimes backfired. Marty McCarry illustrated this once very well, when they had a wood tug on the drydock for replanking. When the job was finished, someone discovered they had planked over the outlet pipe to the toilet. A well-kept secret was the fact that someone had used the toilet, while it was on the drydock. Impatient as always, Ralph said he remembered where the pipe was. He grabbed a spiking maul and swung it against the hull – with predictable results. Bingo! Bullseye! on the first blow. A quick trip to the showers and the dry cleaners completed his "work" for the day. This story triggered Marty's memory about something else in the same vein. The Matton's had a sure-fire way to keep the men from lingering when nature called. Their outhouse was built over the river. The Arctic blasts discouraged a prolonged stay in winter and the scorching sun did the same in summer.

A little about Marty McGarry. He was a part of the Matton "extended" family. His wife was related to Emmett Collins' wife. After some years with the Mohawk Paper Mill, he joined Matton in 1944 as a pipe fitter's helper. History sometimes repeats itself, because his career and Emmett's followed the same path. Both on their own quickly realized their best chance for long employment was to be a skilled welder. Both also on their own, on lunch hours, picked up a welding lead and started practicing. As with Emmett 15 years earlier, Ralph approved. With the help of some older welders, he soon was able to take the test to become a certified welder. The next step was to become welding foreman, and when Emmett left he became Superintendent of the Yard for Bart Turecamo when he bought the Matton company.

In 1928 someone wrote that John E. Matton had the soul of an artist! That would be the last word I would have chosen to describe John E. It came about because Hollywood decided to make a movie about the Barge Canal.

Early in 1928 Universal Pictures Corp. bought the rights to a story by Rubert Hughes called "A Girl on a Barge." The first step was to change the name slightly, to "A Girl on the Barge." It was what they referred to in those days as a melodrama. The setting was barge life in the 1920's on the New York State Barge Canal. The hero, a mate on a tug, was played by Malcolm MacGregor. The heroine, daughter of a drunken barge Captain, was played by Sally O'Neil. The barge Captain was played by Jean Hersholt.

Another actress in the picture, that in later years became a star comedienne, was Nancy Kelly.

The original story itself, not counting Hollywood's normal embellishments, must have upset many real boat people. Neither the original story, nor the movie came close to portraying barge life as it really was, according to all the oldtimers I talked to. But that should come as no surprise to anyone. I quote the Mayor of Whitehall, where the movie premiered, and you will see why the old timers gagged at what he said: "It is a simple, true to life story of canaling." Further he said, "Edward Sloman (the Director) has put on the picture, *without a slip* in his portrayal of true life."

The attempt to film it in Hollywood was a dismal failure. An article in the *Universal Weekly* (house organ for Universal Pictures) of June 2, 1928, provides the details. "The Erie Canal has to have a double. This is not just because the Erie Canal is not just outside of Universal City either. When Carl Laemmle bought Rubert Hughes' 'The Girl on a Barge,' the matter of locations was entirely subordinated to the romance and powerful nature of the story. But as the screen treatment grew, it became more and more apparent it would be necessary for Edward Sloman to make this picture on the original locations, which were specified by Rubert Hughes."

In those early years, location trips were the last resort of movie companies. Photography required expensive paraphernalia. "The shipping of electric-light trucks, sun-light arcs, wind machines, etc. was difficult and expensive. Relaxation of the discipline of studio management so greatly prolongs the time consumed in making the picture that every location trip is in danger of costing two to three times what it would if made in the studio and the locations built on the lot under the supervision of competent, fast-working production departments."

But all attempts to fake the Erie Canal in or around Universal City just stumped the Production Department and a location trip was finally decided upon. In fact, so sold did the Production Department become on this idea that it was determined to make the entire picture in New York state, using a New York studio for the interiors. Edward Sloman, the director, Jack Voshell and Jackson Rose, the veteran cameramen, came on ahead of the company to pick out the proper locations on the Erie Canal and to secure the proper equipment such as barges, tugs and other properties for use in the picture. They were met in Troy by Arthur Cozine, who had been busy for two weeks getting all the data together on the location proposition.

Now comes the unvarnished truth about the Erie branch of the Barge Canal. A trip through its highly efficient and commercialized waters disclosed the horrid fact that there was nothing pictorial, beautiful or filmable in its whole stretch between Buffalo and Albany, where it joins the majestic Hudson River.

"Why, if we would make a picture," declared Sloman, indignantly, "on this canal, they would swear we faked it in California. It just doesn't look like what people think the Erie Canal looks like."

What to do?

John E. Matton of Waterford came to the rescue. In addition to being a contractor, Matton had the "soul of an artist" and he knew right away where there was a wonderful substitute for the Erie Canal. It was the Champlain Canal [a division of the New York State Barge Canal] and the place selected was at Whitehall, one of the most beautiful spots in the Adirondack-Champlain region. Sloman went, saw and was conquered.

John E. not only had the soul of an artist, he also supplied the tug and barges used in the movie. The tug was the P.C. RONAN; the Captain was Joseph "Shin" Roberts. Movies were apparently shot a lot faster in those days. All the outdoor shooting was to be completed in just two weeks. All the interior work was done at Paramount Studios in Astoria, Long Island. The location headquarters for the film crew was in the Queensbury Hotel in Glens Falls.

This was in the days of the new phenomenon called "talkies." Everyone was not convinced that the sound picture was here to stay so this film and many others at the time were made both as a talkie and a silent film.

I was able to pick up a few sidelights about making the picture from "Shin" Roberts' son Jerry. No one, of course, in the film crew could operate a tug. Shin would do it, but he could not be seen by the camera. He did it by kneeling in front of the wheel, with the actor standing behind him. He could just see enough to steer by and the cameramen picked up the shoulders and head of the actor. It was not an easy job for "Shin."

Shin Roberts also told his son about several fight scenes in the movie and a like amount off camera. According to Jerry's father all was not sweetness and light among the actors and crew on this film. It's difficult to find a copy of this film. Many of the old nitrate films were destroyed because they were very flammable and required special vaults for storage.

1929 was the year the "Great Depression" started. Whole industries collapsed. The depression lasted up to the beginning of World War II. While putting all of Matton's tugs on a graph, I discovered they had run the most tugs (9) during these hard times. The years before and after they averaged 5-6 tugs. The Mattons certainly didn't get rich during these years, but they survived intact. Having the yard and drydock helped, along with their reputation for excellent work and reasonable charges. In addition, they had some salvage and repair work away from the yard. Probably not having all their eggs in one basket also helped them through trying years.

As near as I can determine, nothing of great interest happened between 1929 and 1938. The minutes of the annual meetings for 1938 said that *salaries* of the key executives (John E., Ralph and Margaret) would continue while *absent due to illness or accident.* I felt a little spooky reading them, since less than one year later John E. was to have his first stroke, from which he would never recover enough to return to work. The record also pointed out the fact, that in spite of being in the middle of a depression, they were financially strong enough to adopt this policy. The events of 1939 would have long-lasting affects for the Matton's. On the positive side was Ralph's marriage to Margaret and the building of their first steel tug. On the negative side was John E.'s stroke.

You really couldn't say Ralph and Margaret had a torrid, all-consuming romance. After all, he had known her for 21 years before proposing marriage. It was more like an alliance of respect, devotion to business, with maybe a little tender companionship added. The business was their first love, as attested to by most everyone who knew them. Ralph never seriously dated other women. In any event, their marriage may not have been a question of why? But rather, why not? Because of age and other considerations, they never had children but like many other childless couples, they doted on and spoiled other people's kids such as Emmett and Ethel Collins' son Tom.

Because of his father's stroke, Ralph took over control of the business with his father's blessing. One of Ralph's first moves was to build a steel tug. For years John E. had advocated using steel for building boats. The reasons they didn't start sooner were two-fold. First the Depression was in full swing, with a very small demand for new tugs, and secondly John E. had a strong aversion to going in debt for whatever reason. He tried always to pay each bill on the day it arrived. To borrow a couple of thousand dollars to build a wood barge was one thing, but to borrow $100,000-$300,000 was quite another matter. They both deserve credit for waiting, for the record shows they made few errors in judgement.

So in 1939, the first of four JOHN E's was launched. She was 83 ft. long and carried a 600-horsepower diesel engine. The size of their tugs did not increase very much, but the horsepower did. The last JOHN E.

had a 1800-hp engine. Four tugs named JOHN E was a bit much, I think they could have used more creativity in picking out names for their tugs. Nonetheless, the Mattons did better than a lot of companies that simply used numbers, such as the Russell Bros. firm.

The first few steel tugs they built were constructed in a manner far removed from the normal. Everyone joked that Ralph Matton built the tug and then drew up the plans. Actually, he did design them as he went along, with each section on the mold loft floor. After the first two or three tugs, Ralph hired an old R.P.I. graduate to make decent drawings from his own rough sketches.

I kept looking for a color photo of a Matton tug but never found one. The color of tugs is important because they are a floating advertisement. Matton had lots of black and white photos, but no colored ones. Perhaps this was one of Ralph's "economy measures." When I asked Marty McGarry if Matton ever considered making his tugs as beautiful as Bart Turecamo's boats, he told me Ralph said they didn't need to. "Give it to them clean and strong and get it to hell out of the yard," was Ralph's philosophy. It didn't seem to me that Ralph ever got any thrill, kick or pleasure from setting foot on his boats, other than some trial runs. Marty felt that as soon as the tug was launched, he lost all interest in it. At any rate, Matton tugs had black hulls and buff superstructures, with white trim and black stacks.

There was one more first for Matton in 1939. Even though the war had not started for the United States, movement of material to Europe had increased. As a result, Ralph found a demand for his two steel tugs (the new JOHN E. and the MATTON 20) in New York that winter. From this time on he had boats working in the winter. In 1940, a chance meeting in New York with an official of the government of Peru led Ralph to his first contract to build a tug for someone other than himself. In 1940 and 1941 he built two tugs for Peru – the TIGRE and the CURRARAY.

Matton's ownership of the first JOHN E. was short lived. She was sold to the Navy on December 1940 for $120,500. After delivery to the Brooklyn Navy Yard, she was converted to a net tender, renamed TAMAQUE (YN-52) and operated in Boston harbor for the next four years. She was later classified as a medium harbor tug (YTM-741). As is the custom in the Navy, tugs are given American Indian names. Tamaque was a Delaware Chief of the Unalachtigo tribe during the Eighteenth Century. She was taken out of service at Marginal Wharf, South Boston, on December 20, 1945 and in 1946 sold to a company that after several name changes became the Boston Tow Boat Company and the tug now renamed the ATHENA. The latest records have her registered out of Baltimore as the JAMES A. HARPER.

With the war years upon us now it was "a turning point" for the Mattons. As Frank Bushey once told me, "Hell, anybody can make money during a war." It certainly was true for the Mattons. With cost-plus contracts, they built five tugs for the Army and six sub chasers for the Navy. In addition, they took over a tug contract from a shipyard in Newburg, New York, that had defaulted. Emmett took some welders to Newburg to weld them up just enough so they could be towed to the yard in Cohoes for completion. Besides the boat building, the Mattons kept five tugs and one oil barge running during the war. At the height of the war, there were about 340 people working for them. The normal complement when starting a tug was 25-30 workers, 50-60 near completion.

Ralph's proudest boast was that he could make almost anything needed to build a tug. He did, too. In some cases he might have been better off buying instead of fabricating. A few examples come to mind. Except for the compressor units, he built the entire refrigerator, including the wood doors, which were a credit to the cabinet makers craft. Something else impressed me even more: some of Matton's tugs had hydraulic wheelhouses, so that they could be raised and lowered for the bridges on the canal. They were a copy of the lifts used in garages to raise cars. As in any hoist, the most important part is the lift cylinder, which requires precision machine work to make. It is a specialized field and most people would not attempt to make them, but Ralph did, and they worked perfectly.

Some of their tugs did not have hydraulic wheelhouses. Something would have to be done so they could be used in New York during the winter, where the barges were too high to see over. Ralph solved this problem by building a separate wheelhouse that could be set on top of the existing one. They worked just fine, but as Marty McGarry said, it was a real chore to put them up and take them down because all the controls, linkages, etc., had to be transposed each time. With tongue in cheek, I would have to say his two homemade yard cranes stand as his crowning achievement. They were strange contraptions by any definition. The simplest way to describe the crane would be to call it a derrick precariously perched on top of a railroad box car. First, he lay down rails the length of his yard. They were placed eight feet apart, three feet wider than standard railroad tracks. Ralph needed the extra width to provide stability, that was marginal at best, with his top-heavy monstrosities.

To balance the picture, let's look at two things that didn't turn out well. According to Marty McGarry, Ralph's attempt to build a brake to bend sheet metal was a dismal failure, and he probably spent twice the cost of a new one from a tool company. The other one was more in an experiment to increase the efficiency of a tug's propeller (always called a "wheel" in the industry) in the same way a Kort nozzle does. Briefly, a Kort nozzle is a cylindrical steel shroud built around a propeller which acts as a tunnel to better direct the flow of water through the propeller. This increases the thrust, which is the key to more efficient operation of the tug. What Ralph tried was to wrap a piece of copper sheet around the propeller brazed to the blades themselves. In a Kort nozzle, the blades, of course, don't touch the nozzle, otherwise the tunnel effect would be lost. All Ralph's invention did was churn water and vibrate like hell. But I certainly would give him an A+ for trying. In later years, they did buy rather than fabricate themselves. Maybe he realized he was "penny wise and pound foolish," as they say.

The war was not quite over when Ralph got back to his first love and started a new tug for himself. The second JOHN E. was launched in 1945, and the biggest change from the first JOHN E. was that it doubled the size of the engine to 1200 hp. Throughout the history of the Barge Canal, one finds that while the canal tug didn't change a great deal in size, its engines kept getting larger and larger.

Marty McGarry explained to me how Matton was able to shape, roll and bend heavy sheets of steel to the curved shapes required for the hull. He reduced this problem considerably by building a "chined" hull. Long flat narrow strips formed the arc of the hull. It was not a smooth, true arc, but it is a much simpler way to build. To roll and form smooth, complex curves requires very large expensive machinery. A hull constructed in this manner is called a "molded hull." Using the chined method also saved them from making frames (ribs) in a true arc. The frames were cut from angle iron and welded to form the segments of an arc. One part of the boat was rolled, and that was the wheelhouse where the steel was much thinner. I wondered how the plates were formed to the frames. I assumed they were heated, but I was wrong. Marty McGarry explained that heating was too difficult to control. The plates were formed cold by mechanical means such as chain falls, come-a-longs, wedges, and then welded in place.

One of the more dramatic scenes in boat building is the installation of the engine. (Refer to my photos to get a perspective of the size of them.)

168

For the first few tugs, Ralph would launch the hull and tow it to the Port of Albany and install the engine there. At that time he didn't have a railroad siding into his yard, so the Port was the next logical place to deliver the engine. John Mullins of Waterford provided the mobile cranes to set the engines in the hulls, but this arrangement was less than satisfactory to Ralph. One problem was that Ralph's men were non-union, and the union lost no time in getting to Ralph with their demands. For every man he had on the job, Ralph would have to hire a union man. Being completely frugal, this ultimatum forced him to lose his cool, and his swearing could have been heard in Buffalo. Independent in all decisions, he soon put a stop to that nonsense. He had a siding installed to his yard and ever afterwards installed his engines without the interference of the union.

Ralph had another reason to dislike union. When you went to work for Matton, it was understood that no matter what your trade was, you were expected to turn your hand to anything that needed to be done. Of course, unions are dead set against this, because this reduces the number of men needed. Without such flexibility, plus low wages, Matton could never have survived. Also, up until Emmett Collins made a case for change, everybody was paid the same. Emmett convinced Ralph that it was better business to pay men according to the skills required. Thereafter, welders were worth more than painters, for example.

Pay was adjusted by using a shipyard in Newburg, New York, as a guide. The rule of everyone working more than one trade was flexible to this extent. Whenever possible, a good burner who conserved gas and the best welders were kept at their trades. To speed things up, sometimes the welders were put on piece work, that is, they were paid by the foot of weld. There can be a trade-off when this is done. There is a possibility of "slugging," which is the practice of laying whole welding rods in the seam, to partially fill it up, so that one has a lot less welding to do to close the seam. This results in a weld that is not strong enough, however. It can be detected by x-ray, but that is far too expensive and time-consuming, unless one has a cost-plus government contract. Emmett said they had to take their chances that the men were honest. Reflecting on Emmett's character, I suspect he didn't rely totally on that assumption.

One area that Ralph always subcontracted for was electrical work, the main reason being that it is very specialized in a marine application and, there were only about 8 to 10 weeks of work a year.

I talked with Emmett on the subject of wages and learned that there never were any hard or fast rules as to how they were given. Emmett's first raise came from John E. over Ralph's objections. It was never easy to get raises out of Ralph. During Emmett's time as Superintendent, the men would start grumbling that they were not keeping up with the rest of the world. Emmett would then take their case up to the office and argue for them. If Matton decided it was time to give, they all got the same amount, including Emmett. Merit raises were unheard of in those days. After some years of experience Emmett said he could sense when the men were getting restless. And very shortly, he'd have to make the trek to the office. With Ralph and Margaret's volatile natures, I'm sure it was a task that Emmett dreaded.

There is another point to remember. During most of the life of the Matton yard, management was king. The boss didn't have to have a reason to fire you. There were three men out of work for each one working. If you got laid off, there was no unemployment insurance, no welfare, no benefits. Fear was a great motivator. I can only guess that the old timers who voted out the union somehow felt that even if the union might do something for them, the Matton's would sill have final control of their lives. Some benefits did come in later years but most of them after Bart Turecamo took over in 1964. In the 1930's and 1940's there were no paid vacations or pensions. You were allowed to take time off if someone died, but you sure got your pay docked for it. The rule was that you got paid only for the time punched in and out on the clock. Don't fault the Mattons for this attitude, for all small companies during these years were run in the same fashion. As Marty McGarry said to me, "The attitude toward work and absenteeism today would run Ralph Matton right up a tree."

Yet, in spite of this Ralph was no monster. He could show compassion at times. One winter when they were replacing the sub-base on the engine of the R.E. MATTON, they had four men working in the close quarters of the engine room. One of them was Marty McGarry, who was swinging a 20-pound maul against a wood block, placed next to the sub-base to check for thrust movement. An old machinist, Earl Hamilton was kneeling down to watch the movement. Just as Marty started his next down swing, Earl leaned forward to place his finger where he could feel any movement. Marty saw Earl shift his position out of the corner of his eye, but it was too late to stop his swing. Instead of hitting the block square, the maul hit it a glancing blow, careened off and hit Earl in the head. Marty said it sounded like when you hit a steer in the head in the slaughter house. A sickening hollow, THUNK!, Earl went ass-over-tea-kettle and Marty was sure he had killed him. Earl came around a little as they were getting him into a car to take him to the hospital. Marty doesn't remember how long he stood there. He probably was in a state of shock. The next thing he remembers is the door of the engine room swinging open, and Ralph's bulk filling the opening. In his normal booming voice he said, "Well you big sonofabitch, you did it this time." For a moment Marty's stomach dropped out and his knees turned to Jello. Ralph snapped him out of it when he spoke again. "Don't feel bad about it. I did the same thing once. I hit a guy and almost took all his fingers off. Earl is going to be all right, so go back to work and forget about it." Marty said to me, "I thought that was a hell of a thing for him to do." This was the only machinist we had and I thought I had killed him. He thought enough, not of me, but how I must feel to come down to console me." Ralph could empathize with Marty because he too was a man of great physical strength, one who loved manual labor, yet knew the risks involved. In fact, as many have said, Ralph never was happy about going into the office full time after John E. had his first stroke. He was in his true element only when he was in the yard working alongside his men.

In 1946 they brought out a new tug named for his wife Margaret. This was also the first year Matton had to borrow a quarter of a million dollars to build a tug. This one he named after his Mother Edna. In 1948 they started a project that would turn out to be one of a kind for them. They received a contract from Oil Transfer of New York for an oil barge. It was the biggest thing they ever built 210 ft. × 43 ft. × 15 ft. Launched in 1949, it made quite a contrast to the smallest vessels Matton ever made – three tiny rowboats for the Little Sisters of the Poor, a religious order.

Up until 1948, even though he took no active part in the business, John E. would be driven to the yard every day for an hour or two. Tragically, that all ended that year when he suffered another stroke. He went back to Leonard Hospital in Lansingburgh (North Troy) and, except for special occasions when the family would take him home for a couple of hours, he never left the hospital until his death in 1959.

From what the records show, 1948 was not a good year. Due to conditions in the industry, the treasurer (Ralph) had to pay out of his own pocket all traveling and entertaining expenses for the coming year. Besides launching the Oil Transfer barge in 1949, Margaret was invested as Secretary and Vice-President of all the companies, except the original. This meant that Ralph's mother was out as an officer, but she had been a figurehead only, so I doubt if it mattered to her. Edna Matton's life revolved around her home and family. In 1949 John E. died, and the next year saw another trip to the bank for a quarter of a million dollars to start the EDWARD MATTON (named after Ralph's grandfather). In 1951 profits turned out so good that the company voted themselves a 10% bonus and reimbursed Ralph for his out-of-the-pocket expenses.

The work Matton's did from 1952 through 1954 was the result of another war, what President Truman called the Korean "police action!" In any event, Matton built 10 small (100 ft. × 40 ft. × 12 ft.) barges. Half were for oil, the rest for deck cargo. In addition, they built two tugs for the Air Force, the first time Matton built two boats at one time.

In 1957 Ralph got a call in June from the U.S. Navy saying that they had a bomber down in Lake Champlain and asking if he would raise it for them. In most cases like this, the concern is working in too-deep water. Here was the opposite: very shallow water. The airplane was in about nine feet of water, which posed, a problem getting a tug to the scene. Emmett took four men up to make a survey. They reached the plane, part way by ferry boat and the rest in a rowboat. After returning to the yard they explained to Ralph what was needed. Ralph immediately called Oil Transfer in New York and rented one of their oil barges. As soon as it arrived at his yard they stripped most of the machinery and pipes off the deck. He needed enough clear deck space to carry one of John Mullins' crawler cranes and the plane. Putting that huge crane on the deck might prove to be a real chore. No one had equipment big enough to lift it on. They lucked out in that the cement wall (State Dock) next to his yard was just the right height, so they could drive the crane on to the deck. They lightened up their tug, the H. A. MELDRUM, so she would draw no more than 9 ft, and headed up the Champlain canal to the lake. It wasn't possible to raise the bomber in one piece, so divers were sent down to rig cables and cut the tail loose. It took about three weeks to complete the work. Matton was paid by the day for men and machinery. Looking at the photos, I would guess that since it was a job for the government, all the payrolls were padded just a tad. The reason for the crash? It was alleged the pilot "forgot" to switch fuel tanks and ran out of gas. At a high altitude this would not be a problem, but it appears he was making an approach to Plattsburg Air Force Base. If this were the case, he wouldn't have had time to make the switch.

Ralph was not an armchair boss. He knew exactly how long a job should take and you couldn't bluff him. He had a short fuse, and he would bawl you out at the drop of a hat. As a matter of fact, his hat. One of his actions when angry was to slam his hat on the ground and stomp on it. He always wore a fedora to hide his baldness. Marty McGarry described Ralph in a different fashion. He said Ralph was a man of the men. He would listen to you no matter what job you might hold, laborer or top mechanic. He might agree, but if he didn't, he would tell you, as Marty said, in very "broad terms." He was at his best with a tool, not a pencil, in his hand.

I once asked Emmett for an example of what would set Ralph off. Emmett and a crew were switching two new towing cables from one tug to another. Instead of taking the time to rig phones, they used hand signals with disastrous results. One tug didn't stop when it should have, so the line tied to the cable snapped, dropping one cable in the Hudson River. Mind, that cable was worth $5000! Captain Jimmy Clinton offered to go with Emmett to Ralph, but Emmett refused. Emmett called Ralph and asked if they had any hooks that might work to retrieve the cable. Holding the phone as far as possible from his ear, he had no trouble hearing Ralph's answer.

The language should have melted the wires. After some difficulty they managed to save the cable and put it back on the spool. Again, believing the old adage that discretion is the better part of valor, Emmett called instead of going to the office, to give Ralph the good news. Ralph made only one comment: "You're a lucky sonofabitch."

I asked if Ralph had any hobbies. Most of us have our minds programmed to expect a particular answer to certain questions. In this case, I expected the usual, such as golf, fishing, etc. I received a delightful surprise when Emmett said, "Ralph liked to go out and look the girls over, go to the movies alone once a week and get an ice cream soda." Naturally, this was before he married Margaret. In today's world, he would have been called a "square." Marriage didn't change things very much. He and his wife were not night people nor did they entertain much.

There was a small amount of business entertainment. When bringing out a new tug, all the officials and guests on the trial run would be served sandwiches, coffee and drinks aboard. After the trials, they would be taken to either Keeler's or Duncan's in Albany for a good meal. As a matter of interest, even after John E. was impaired by his strokes, he would sometimes snap out of it and mention these restaurants to his nurses.

Ralph liked big cars but like many in his business, he never owned a pleasure boat. He was conservative in dress as well as politics, a strong church-going Catholic and always generous to the church. When Sister St. John came to the office there was always a twenty dollar bill waiting for her on the edge of the desk. He and Margaret were dropping in twenty dollar bills on Sundays when most folks could only afford nickels and dimes.

I tried to get Emmett to comment on Ralph's opinion of his competitors. He refused with one exception: the Albany harbor plum. The Albany harbor tug is used to dock and undock large ships. It may come as a surprise to some, but Albany is a very busy harbor. The Coast Guard ice breakers keep the channel open in the winter, so it has a year-round harbor. My friend Jack Maloney has turned around as many as 80 ships in a busy month. It can be a tricky job. For example, the cement ships that unload at Ravena (just below Albany) can not be turned around there; they have to be brought up to the turning basin at Albany. When they are broadside to the river they just barely clear each shore. With an ebb tide running, the ship acts like a floating dam. I can appreciate the skill required to turn these monsters around and head them downstream.

In the beginning everybody wanted this contract, so a degree of under-cutting fees went on. About the time of World War II, Ralph put an end to the battle and said he would turn the ships around for nothing. That was enough for the New York outfits. They backed off and Ralph had it all to himself, and it was worth the fight. Year after year it was the only sure money-making tug they owned.

As the business improved Ralph and Margaret started taking vacations. They followed his Mother and Father's footsteps by going to Florida and taking cruises to the Caribbean. There was one big difference, though, for Ralph or Margaret would call the yard every day for a full report from Emmett. Emmett saw nothing wrong in this behavior, although it was more like parents checking on their child while they were away. The business, in a sense, was the child they never had. They never had to break a vacation to come early, because Emmett kept things moving right along. As a matter of fact, some people say that things moved a little more briskly when Emmett was in charge. For the most part Ralph and Emmett kept their social and business life separate, which must have been difficult at times.

They would have thunderous arguments and Ralph would threaten to fire Emmett, and he wasn't just "blowing smoke." What were the arguments over? Cost estimates of jobs. Each one would make out a bid sheet and then Ralph would compare them. The bottom line was that Emmett's were usually higher than Ralph's, because he had this silly notion that you should make a decent profit on your labor. Ralph was content to break even at times, so that the yard would not be empty and he would not have to lay off his best workers.

They were both right, but I lean a little toward Emmett's view. Ralph probably could have raised his bids and still got the contracts. I doubt if his competition ever worked just to break even. I certainly can't imagine Frank Bushey or Tom and Bernie Feeney doing it.

Ralph had an interesting way of paying overtime. Double time for Sundays and triple time for holidays, but only when they worked on outside boats where he could charge the owners. He wouldn't pay that kind of money for work done on his own boats.

Most of my questions about Margaret were addressed to Emmett's wife, Ethel. Since Ethel's sisters were much older, Emmett's wife felt closer to Margaret and they became fast friends. As with the men, the two women kept business and social life separate. If Ethel or Emmett wanted something from the yard, such as a gallon of paint or some lumber, John E. or Ralph would be inclined to give it to them. Margaret said, "NO, they pay for it!" It was at cost, but it was taken out of Emmett's next pay check. Yet, if an employee had a financial problem, he would go to Margaret rather than Ralph or John E.

When talking with people about the business I picked up what I always refer to as "Legend, Lore and Lies." Though many stories one hears turn out to be false, one exception was the story that Ralph built the tugs and Margaret ran them and the crews. For the most part, this story was true. Although Ralph was aware of everything that was going on, Margaret was the one who acted as dispatcher and handled all the day-to-day operations of the tugs. She also did most of the hiring of the crews. Margaret was very good at these jobs and this was years before any women's lib movements. She could be volatile, especially about money matters, but she was not quite up to Ralph in the salty language department.

In February 1958 they sold the second JOHN E. to the Port Everglades (Fla.) Towing Company for $175,000. This sale led to a contract to build two tugs for that company. The EVERGLADES came out the next year and sold for $410,960. The CHALLENGER was launched in 1960, and I haven't the foggiest how much they got for her.

January 19, 1959 was a black day for the Matton family. John E. died after spending 11 years in the hospital. He would have been 80 on April 2. For all those years he had round-the-clock private nursing. Room 131 had become his real home. Although he went home for a few hours on special days, he was always anxious to get back to the security of his room. Like many long-term patients, he was a favorite with the staff. They visited every day and he was known as Pop to all. He had regressed to a boy of mid-teens, although at times adult thoughts would surface for short periods of time.

He had many nurses during those 11 years, and I was able to talk to two of them. Helen Stallmer (now Mrs. William Pearson), who was John E's nurse for five of those years, and Grace Hancox, the Mother of my next door childhood buddy. I knew Grace had taken care of John E., so I went to her first and she led me to Helen. Grace had given up full time nursing, but substituted for any of the nurses that wanted a day off. Time off for private duty nurses was hard to come by. They were hired to work seven days a week. The families did not like to see substitutes. This again, as I have shown before, is another example of the prevailing attitude of the rich toward the working stiffs of the world. Because of the depression they had it all their way.

As an aside to this section about John E. there was a startling similarity between Grace and Helen. Both were strong, earthy women, with great compassion for the sick, the way we all hope nurses are. They used the same expressions and their voices were so much alike you would swear they were sisters. Grace Hancox was very special to me. My Mother died when I was 12 and she became a second Mother to me.

I've digressed long enough. The one thread that runs through any conversation about John E. was the devotion of the family. Except for short vacations, Ralph, Margaret and Edna visited him every day for 11 years! Most people raise their eyebrows when I tell them this. I find it difficult to accept this behavior myself. Could there be another reason for this intense devotion? One theory says it may have been for appearance, because of their position in the community. If all the skeptics, including myself, are wrong, then these people are certainly candidates for canonization. Remember, John E. had regressed to a point where adult conversation was all but impossible, and they kept this vigil for 11 years! So, just maybe, once again, truth is stranger than fiction.

After Helen had been with John E. for a few months, he had improved enough to be understood and alert enough to ask for things. One thing he asked for was a cocktail. The Doctor approved, so Helen would put about a ½ oz. of scotch in 3 oz. of 7-UP. Not a very potent drink, but

John E. looked forward to it each afternoon. Helen wasn't sure he really enjoyed it, but it was something he had focused on from the past. Well-l-l, it didn't take Edna but three days to hear about it and react in her usual demure fashion. Quote Helen, "Mrs. called me at home and bawled hell out of me, when she should have yelled at the Doctor." But no, you yell at servants such as a nurse, not at a professional such as a Doctor. Incidentally, Grace and Helen used the expression, "Mrs, said this," or "Mr. said that," and always left off the last name. I don't know why except in some Irish families they would speak of their spouses as Mr. or Mrs. instead of using first names. The bottom line is that John E. kept getting his cocktail and Helen Stallmer kept paying for it out of her own pocket.

In his condition, John E. had few pleasures, but one of them was going for a ride in the afternoon when the weather was nice. Helen would call the yard and ask if the car was available and if so, Hazard (Isn't that a great name!) Wheeler, the chauffeur, would take them for an hour's ride before supper. Some times the car would be promised and not arrive because it was vetoed by Edna. When this happened, John E. would get very upset and it would take a long time for Helen to calm him down. The tantrums were typical of a frustrated young boy, which is exactly what John E. was at that time. Once to calm him down Helen got him dressed (no easy chore), put him in a wheel chair and pushed him four blocks down and four blocks back to the hospital. By the time she got John E. back to his bed she was exhausted and in just the right mood to read the riot act to the family, which she did. It cleared the air and after that when the car was promised, it arrived.

With the exception of the daily visit by a priest, John E. had no visitors but the family. In those days it was possible to stay years in a hospital. Today he would have been in a nursing home. With round-the-clock nursing he could have been home, but that would have put a terrible burden on his wife. Plus Ralph had said that as long as he had the money, his father would have the very best care the hospital could provide.

And the costs were staggering. It averaged about $25,000 a year, for a total of over $250,000. With long term patients, families sometimes give the nurses some monetary gift. *All* were "positive" the Matton's did. No one got one penny for all their devotion to John E. They weren't looking for anything, but it would have been a nice gesture. I hate to keep busting everyone's bubble, but the fantastic stories about the Matton generosity never proved out.

John E. and Ralph were alike in some things. Both had a fondness for good food. They liked Florida, were conservative in dress, except John E's golf togs. John E. had one small affectation. He smoked his cigarettes using a holder, a la President F.D. Roosevelt. Unlike Ralph, he rarely went into the yard. He spent his time in the office and no one can ever remember him setting foot on one of his tugs.

Information about John E's wife was sparse. After the first few years of struggle to get the business established, she stayed in the background. I asked Ethel Collins to describe her. Some people would have taken a few moments for reflection, but not Ethel. Her answer popped out like a squeezed pea from a pod – "She had a house cleaner than a hospital, and she was a good cook."

Once in the hospital, after a trying episode with Edna, Helen asked John E. why the hell he ever married her. His answer: "She was a good cook." Helen pressed a little reminding him about Jessie, a girl he liked to go ice skating with. Yes, she was nice, but Jessie's skating couldn't match Edna's cooking.

Edna was a neat freak and a frugal one at that! Anything that needed done around the house was done by the men from the boat yard. They were not thrilled with these assignments. One job they had every fall was putting up the storm windows. She supervised every move and insisted they wear white gloves while doing it. Another job they hated was the weekly ritual (and it was a ritual!) of washing the car. The drill went something like this:

Step 1. Stand by the car with hose at the ready until Mrs. turned on the water from the house. You certainly couldn't expect those men to respect her passion for conserving an expensive commodity like water.

Step 2. Wet the car down. It was wet enough when Mr. shut the water off.

Step 3. Walk to the house, where Mrs. would hand them a damp rag already impregnated with the precise amount of soap. Proceed to scrub down the car.

Step 4. Wait for the water to be turned back on and rinse the car off. Return the rag, so it could be washed out for the next time.

What I have described is a typical Navy shower, where there is a real need to conserve water. One man figured out a way to get out of this lousy job. He conveniently forgot to roll up a couple of windows with predictable results. Of course he suffered a tongue lashing, but on the other hand, he was never sent to wash the car again. You may think I'm being hard on Edna. I did not create her character. I will compensate by adding an editorial comment. For Edna Matton, the so-called Golden Years turned out to be solid brass. Her life after John E. entered the hospital the second time became a barren mockery of what we all expect life to be in our later years. We must give her high marks for the daily visits to the man/boy she no longer was able to share anything with. As if to confirm that her life revolved around her family, she died just six days after her son Ralph on September 24, 1963.

To kind of even the score, I'll give you a vignette about John E. At times he could be as frugal as his wife. As a small business venture outside of the boatyard, he built three houses in Waterford – one to live in and two to rent. He saved money by hauling the bricks himself up the Hudson River. Part way through the construction he decided the brick layers were getting too much money His confrontation with them led to a strike. So what was the big deal, he would send men from the yard and let them finish the job. After all, didn't Matton have the reputation that they could build anything? That was not the brightest decision John E. ever made. His men botched the job so badly he had to hire the brick layers back. God, it must have been difficult for him to eat crow. To this day you can see the goof on one of the houses. While I'm on the subject of these houses, let me bury another fairy tale about the Matton's. Once again legend had it the John E. let some of his retired employees live in the houses rent free. Wrong. In fact, in 1959 when John E. died they were collecting $164 a month rent for each of them.

The year 1961 saw a change in where they borrowed money. Ralph dropped General Motors in favor of Fairbanks-Morse for the usual $300,000 to build the second MARGARET MATTON, and installed one of their engines.

The second MARGARET MATTON was built in 1961, but by then the end of the Matton enterprises was in sight. A clue could be found in the company records. There was interest of $6,285.00 due one of their companies from another. They had to default. Looking at their books for that year I saw that they had lost money.

The MARGARET MATTON was launched in 1962 and they went back to Fairbanks-Morse for what would be their last loan to build a tug. Ralph died suddenly at 11 p.m. on the night of September 18, 1963. There is a story connected with Ralph's death that has been used as a theme in a few movie and TV scripts. He had drawn up a new will and wanted his old friend and marine lawyer in New York to check it over before he signed it. Why didn't he have his local, longtime lawyer Edward Pattison look at it? In the early evening of the 18th, he asked Emmett to mail it because he was not going out anymore that night. A few hours later he suffered his fatal heart attack. The new will gave 50 per cent of the business to Emmett, 30 per cent to his son Thomas and the remainder to Margaret. He had said that Margaret had enough money in her own right to take care of herself. A will written in 1940 had left all to Margaret. When the contents of the new will came to light Emmett was truly shocked. He had no inkling that he and his son were to receive the lion's share of the business.

From Ralph's death on, the business continued to slide. Margaret lost interest not only in the business but in taking care of herself. She started seriously looking for a buyer. During this time Emmett wanted to continue bidding on work but Margaret objected, screaming "You want to spend me into the poor house." Emmett felt if the yard was busy she would have a better chance of getting a higher price for the business, because it is very difficult to sell an empty yard. While the physical assets can be duplicated almost anywhere, the real value in a shipyard is its skilled workers. The last (4th) JOHN E. MATTON should have been launched in 1963, but they decided to hold it over and bring it out in the spring of 1964. Emmett took the skeleton of that tug, ordered all the machinery and completed it by himself. It was fitting: T. Emmett Collins from laborer to tug builder.

Before selling the yard, Margaret offered it to Emmett but he declined. His age (60) was one factor, another was his son's complete lack of interest in the business. And finally he was reluctant to risk his savings.

The boats and yard were sold in the fall of 1964 to Bart Turecamo of Oyster Bay, New York. While waiting for the sale of the business, the men in the yard had nothing to do. Marty McGarry told of make-work jobs like resetting the ballast in the railroad bed, cutting grass and painting. Even though they had reduced the crew to about 30 men, Margaret no doubt hated to see money paid out for so little in return.

Bart Turecamo was another Horatio Alger story. His buying out the Matton's didn't surprise anyone in the business. Matton tugs had worked for Bart for a long time in the winters. Turecamo gave Margaret a certain amount down and agreed to pay off the rest in 10 years at 4 per cent interest. Emmett voiced a real objection to the 4 per cent, trying to convince Margaret it was too low, but she was anxious to sell. Here are the results of the sale:

Buildings and Property	$	25,000
Materials & Equipment		75,000
	$100,000	
MATTON 25	$150,000	
MATTON	200,000	
RALPH E. MATTON	200,000	
MARGARET MATTON	300,000	
JOHN E. MATTON	368,874.76	
	$1,318,874.76	

The reason for this odd number, according to Emmett, was the balance on a loan from Fairbanks-Morse. As can be seen, the yard went for a paltry $25,000, which is understandable when the age of the buildings is considered. There was one tug Turecamo didn't want – she was old, worn out, and had a wood hull. They had kept her longer than any other – 33 years. This was the H.A. MELDRUM, which was sold to the Parrott Bros. for $12,500.

I almost left out one boat of a different type they built just one of. My good friend Captain Bob Gordon of the ERIN KEHOE read a book on the Lake Champlain ferries and saw a reference to Matton. I checked it out and, sure enough, John E. had built the ferry CHAMPLAIN in 1922. She was sold to Flora D. Cunningham of Fort Ticonderoga, New York, and was out of documentation in 1947.

For the transfer to Turecamo, the name Matton was to stay on the shipyard, and Emmett agreed to stay for 15 months after the takeover. A somber meeting took place in the living room of Margaret's home on December 28, 1964. Present besides her was Emmett and their accountant Curtis Dubois. They went through the motions required by law to dissolve all the companies. After some 64 years another part of New York State Barge Canal history came to an end. The postscript would be written in 10 months, when, after a brief illness, Margaret died on October 17, 1965.

I shall put to rest the last of the Matton legends. Many people I interviewed were positive that the Mattons took care of at least some of their old employees. High on the list were Captains Jimmy Clinton and Jack Maloney, but they received nothing and didn't know of anybody who did. The same holds true for the old timers who worked in the yard. Though the stories persist, there's no proof of any legacy from the Mattons, except for one person other than family who was remembered. It was Margaret's housekeeper, Helen Holm. The estates break down as follows:

John E. left	$1,787,489.65	in 1959
Ralph E. left	$1,162,280.45	in 1963
Edna left	$ 249,439.04	in 1963
Margaret left	$1,022,175.30	in 1965

There is something important to consider as one looks at these numbers. Most of these dollars, except for Margaret's, were paper dollars. The dollars were based-on the appraisal of the assets of all the companies for tax purposes. There were not many real dollars until the business was sold. I found an example in going over the estate papers of John E. In the beginning of this chapter, we saw that in 1922 when John E. incorporated he had cash assets of $585.94. At the time of his death, 37 years later, it was necessary to borrow money to satisfy the tax collector. The estate papers reported "Balance of cash remaining in the hands of the executors $625.05." In 37 years John E. Matton's net worth had increased by $40.11!

Lawsuits are the one facet I haven't talked about. Although you can have the equivalent of a "fender bender" in the towing business, they are not settled in the same way. Insurance claims and lawsuits are an occupational hazard in this business. You just have to kiss someone's barge or pier and they scream like a wounded rhino. I can't speak for all, but the way Matton handled the curse was by retaining two sets of lawyers. One was local and the other – the marine expert – was in New York. I really don't know why they needed the local lawyer for this type of problem. In addition there were two insurance companies in the U.S.A., plus Lloyds of London, and the Marine Surveyors. Of course the other side had as many or more lawyers.

Once with Ralph and once with Margaret they decided to settle directly with the other party. It didn't take the lawyers long to react to that in a very negative way and warn the Matton's about this kind of behavior. After all, you have to nip this kind of nonsense in the bud. You don't want folks to get the idea they can get things done without lawyers. There are times when it is better to use them, but *every time?*

As I read the cases over, it seemed to be a game of exaggeration, bluffing, and threats. Most times they settled for far less than they originally asked for. And no matter which way it turned out, there was always a guaranteed winner – the lawyers.

This sample case, as you will note, had a bizarre ending. It started on January 13, 1963 and was not resolved until March 4, 1968. Just like you and me, they carried a deductable policy, but the amounts were greater. In the case of Matton's it was $2500.

MAHAR & MASON
COUNSELLORS AT LAW
70 Pine Street
New York 4, N.Y.

March 1st, 1967

Powell & Minnock Brick Works Inc.
vs.
The Callanan Road Improvement Company
Tug CALLANAN No. 1
Tug MATTON and scow IDA
January 13th, 1963
File No. 3780

Messrs. Johnson & Higgins,
63 Wall Street,
New York, N.Y. 10005

Attention: Mr. Hicks

Gentlemen:

We refer to this matter and our letter to you of January 26th, 1966 by which we reported the receipt of a libel naming the tug MATTON, in a suit for loss of brick from the scow IDA. The claim was stated in the sum of $10,000.00. There have been developments which deserve a report to underwriters. The master of the tug MATTON and later its mate attended at our office and their extended statements were taken. From these it appears that the circumstances were as follows:

On the early morning of January 13th, 1966 the tug MATTON arrived at Kingston Point with orders to assist the tug CALLANAN No. 1 with a tow of five loaded scows in heavy ice conditions which were then existent in the Hudson River. The scows were in tandem with the tug CALLANAN No. 1 ahead. The MATTON broke ice forward of the flotilla for about a half hour but as the CALLANAN had difficulty in progressing the tow, the positions of the two tugs shifted. Thereupon, the MATTON took the hawsers from the tow and the CALLANAN went forward to break ice. At this time the mate of the MATTON, who was on watch, noted that the head scow, named IDA, loaded with brick, and which scow was owned by the Callanan Road Improvement Company (whose subsidiary owned the CALLANAN No. 1), was listing to starboard. The list was brought to the atten-

183

tion of of the mate of the CALLANAN who stated to the MATTON's mate that the list was not unusual for that scow. It was then that the towage bridle was put out from the MATTON and the towage continued. The list increased and the MATTON's mate called her master. Thereupon the tow was stopped and the CALLANAN No. came back to the tow. Her master was informed that the MATTON would no longer tow the scow and would have no responsibility for her. The master of the CALLANAN telephoned his office and thereafter two crewmen of the CALLANAN went aboard the IDA, which was unmanned, but found no water in her hold.

At 6:00 a.m. the mate went off watch and the master of the MATTON then was on duty. Fearing that the IDA might dump part of her load of bricks and possibly capsize toward the opposite side and damage other scows in the tow, he decided to turn the tow around to have the IDA as the last scow. He telephoned the master of the CALLANAN over the radio and informed him of his decision. The CALLANAN No. 1 then returned to the tow, made fast to the tow and the towage bridle then was put out from the MATTON to the last scow. The master of the CALLANAN reported over the radio-telephone to his owner that he thought the scow was overloaded by the brick yard, the plaintiff in this suit. The master of the MATTON also had the opinion that the scow was overloaded and stated that he would not continue with her until some of the cargo had been removed. At about noontime on January 13th a foreman and three men from the plaintiff's brickyard arrived and, after the master of the MATTON refused to proceed to New York with the scow alone, the foreman requested that the scow be towed out in the river so that *some brick could be thrown over her side by his men.* At noon the MATTON's mate again came on duty and it was during this watch that the scow was moved out into the river and a quantity of brick cargo was jettisoned. It is said that no cargo was lost during the towage but that the quantity unaccounted for was that thrown over by the plaintiff's employees as described.

Under these circumstances it is our recommendation that, pending further developments, no suggestion of settlement be entertained by the MATTON interests and that the matter proceed to trial. Will you kindly place our letter before interested underwriters and the assured for approval of our recommendations.

Very truly yours,

SMITH, PATTISON, SAMPSON & JONES
ATTORNEYS AND COUNSELORS AT LAW
22 First St., Troy, N.Y.

March 4, 1968

Mr. T. Emmett Collins
414 Third Avenue, North
Troy, N.Y.

Mr. William M. Connors,
Trust Officer of Marine
Midland National Bank
Fourth & Grand Streets
Troy, N.Y.

    RE MARGARET MATTON ESTATE
        ADMIRALTY FILE #3780
        POWELL & MINNOCK BRICK WORKS
        vs.
        TUG MATTON

Dear Emmett and Bill:

One copy of this letter will go to each of you, and to each of you will go a copy of the letter of February 29, 1968, from McHugh & Leonard.

I recommend that we follow the advice of McHugh & Leonard, who have assumed the defense of this proceeding, and if you agree will you please send me a check from the estate for $2,125.00 payable to Hill, Rivkins, Warburton, McGowan & Carey, as Attorneys for Powell & Minnock and another check for $295.94 payable to Mahar & Mason.

EHP:LMM   Very truly yours,

Edward H. Pattison

The mystery question is, of course, why should Matton's have to pay anything when it was the Brickyard crew that threw the bricks overboard? Further, with the scow tied up to the Brickyard pier, why take it out in the river instead of off-loading on the pier, thus saving the bricks? Strange!

Before completing the Matton story let's consider how things were after Bart Turecamo bought Margaret out. Did he carry on the tradition of excellent workmanship of the Mattons, and how did he treat the workers? Bart Turecamo gets an A+ on both counts. The wages and benefits far

185

outstripped anything the Mattons might have done. The biggest change was in the design of his stock tug. The Matton stock boat was single-screwed and 85 feet long. Turecamo tugs are twin screwed and 105 feet long. Matton tugs were designed primarily for canal use, Turecamo tugs were not.

The other, more obvious difference, which Ralph would have rejected was the finish of Turecamo tugs. They are air conditioned throughout with no exposed wiring, and have plush quarters for the crew and a wheelhouse with all the latest navigational aids. The *piece de resistance* is the outside superstructure which is grained to look like wood. It takes an artist to do a good graining job. Turecamo tugs have no peers when it comes to graining.

Marty McCarry said that Bart Turecamo built steadily. From the takeover from Margaret Matton until 1978 the Turecamo firm built four 52-foot police boats and one pollution control boat. The company built four tugs for themselves, and, as a testimonial to the quality of their boats, Turecamo was the builder of the MOBIL 1, the TEXACO TAMPA, and Texaco's LARGO REMO. So all I can say is, "Rest easy, John E., Margaret, and Ralph – you would be proud."

CHAPTER 11

THE BUSHEYS OF BROOKLYN

Like another New York marine dynasty (Moran), the Busheys had their beginning in upstate New York and nearly got sidetracked forever. Although the family came from Quebec, our story begins when the Bushey family moved to Oswego, New York.

Like Michael Moran, founder of the Moran empire, young Ira S. Bushey drove mules on the Erie Canal. No one is sure how long he held this job, but his next action was to heed the words of Horace Greely and head West to gain fame and fortune. Neither happened, but for the rest of his life Ira retained a love of the outdoors, hunting, and fishing.

The next location to which I was able to trace Ira was Jersey City, New Jersey, about 1895. By now he was in the boat repair business. After that, he moved to Staten Island, New York, about 1900; and Ira's final move was made to Brooklyn in 1905.

By 1905, Ira also had a wife, four sons, and one daughter. The first child born was Francis on November 27, 1888, followed by Raymond on March 19, 1891, Ira Jr. on November 14, 1897, and William on February 4, 1899. I don't know when daughter Ella was born or died.

As each boy became old enough, he went to work in the shipyard. The business prospered so that in 1903 Ira S. Bushey and Sons incorporated and Ira made their first expansion in 1917 when he bought the Downing and Lawrence shipyard and gained 1,400 foot of frontage at the foot of Court Street.

Information about the early years of Bushey's company is very sketchy at best. Most of what I learned was gleaned from the pages of the Brooklyn *Daily Eagle* and a small amount from Ira's grandson, Frank.

One thing I hoped to find was a sage piece of advice Ira might have passed on to his sons. He did have a philosophy which at first blush sounded reasonable. It's fortunate that his sons chose not to follow it or this chapter would not have been written. His philosophy was simple and straightforward: "You should not compete with your customers." In other words, if you are in business to build and repair boats, you do not operate them. I have no records before 1922, but at least by then the sons were operating tugs and barges.

The year 1917 saw the first court action against the Busheys that I could discover. The controversy centered around the railroad flats [used to carry railroad cars across the Hudson River] Bushey was building and launching into Gowanus Canal. The complaint came from the Tebo Yacht Basin where it was felt the launchings might damage one particular yacht placed in their care. It was the CORSAIR owned by multimillionaire J. P. Morgan. When the case came to court, Judge Kelly apparently felt the Busheys knew what they were doing because he denied the injunction with these words, "History of business shows accident unlikely." He further suggested that if they were worried, they could move the yacht during launching.

By 1919 the Busheys had built 254 barges, railroad flat and scows. To say that 1920 was a rather eventful year for the Bushey family would be an understatement. The month of May must have been memorable, to say the least. The bad news started on May 13 when Ira's youngest son, William – then age 21 – was arrested after a motor chase. William bet his best friend, John Randolph, that he could not sit on top of his car on a short trip. This was a result of a party they had attended. William was charged with reckless driving and disorderly conduct. They were locked up in the 30th Street station. The disorderly conduct charges were dropped, and the case was sent to traffic court, a place with which William would become very familiar. What makes this anecdote special is the action of the police who made the arrest. In their zeal, they shot at these dangerous criminals and grazed Randolph.

Things looked a little better on May 20 when the Bushey Company completed a new 15,000 ton drydock that could lift a ship in just 18 minutes. But then, there was bad news, for Ira Jr. was accused of assaulting an employee of their shipyard. It seems as though a pipefitter named Walter Madden was not working as fast as Ira thought he should and he showed his displeasure by punching Mr. Madden in the eye. As usual, I could find no follow up in the newspapers as to the disposition of the charges. The last of the bad news happened in August when the shipyard went on strike.

A final news clip of 1920 tells us that Ira S. had set sail just before Christmas for Europe and Egypt. He had an audience with the Pope and found time to visit the tombs of the Pharaohs. I have a feeling that Ira may have said to himself, "I've had it with this family, I'm going to take a trip."

William's escapades with his car continued, but things came to a head on December 30, 1921, when he was fined $100. He offered the revocation of his license and promised the court that he would not drive for five years. Previously he had been fined five times for speeding and reckless driving. The cost in his short career was $550 – a huge sum of money in 1921.

Now that Ira's sons had convinced him they should be in the transportation business as well as the shipyard, a new company was incorporated on March 27, 1924. A man about whom little is known became a partner. His name was Bullock, and the company was called the Brooklyn and Buffalo Navigation Co. Their boats were named B & B #1, etc. From this, many people assumed that the B & B stood for Bushey and Bullock, but of course it didn't. B & B handled dry cargo between Buffalo and Brooklyn. In its hey day, they were pushing over 30 barges in this service.

A decision in the mid-1920s pushed Bushey into the big time. The family decided to invest heavily in the fuel transport business. The number of automobiles was increasing at a fast pace, plus the conversion from coal to oil for home heating was just beginning. As a result of this, a new company was formed on November 14, 1925. It was founded with two partners and Bushey and was called Spentonbush Fuel Transport Service. The name came from the three partners: *Spencer, Toner,* and *Bushey.* In time it would become the key operating company. In all, Bushey would incorporate over 30 companies, but when people speak of Bushey, they rarely remember the others but always speak of "Spentonbush" as if it were the only one. Unfortunately, the founder, Ira S. never lived to see the birth of this dominant company. He died at home on September 20, 1925. He worked up to the day he died and his estate was estimated at $250,000. Frank said their method of operating would give today's government regulatory agencies fits. They rigged some steel tanks on the decks of scows and pumped fuel from them to trucks on shore.

One more important company, Patchogue Oil Terminals, was added in 1931. This company was responsible for the storage and distribution of fuel from both their Patchogue, Long Island location and the terminal next to the shipyard. Bushey had now covered all the bases. The Ira S. Bushey and Sons shipyard built and maintained the fleet. Spentonbush was responsible for sales of product and the operation of the fleet, and finally Patchogue Terminals the storage and distribution of products on shore.

In Frank Bushey's view, the reason they have been successful is because they were in the "transport business," meaning owning tugs, barges, terminals and hauling your own products, as opposed to general towing. He could be right if we remember that most of the companies that were engaged in general hauling are gone. I think there is another factor to consider. The Busheys had the foresight to see the end of hauling dry cargoes and had the courage to switch to the fuel market before their competitors saw the potential of making this change.

Having committed themselves fully to liquid cargoes, they added another type of vessel to their fleet. In the 1930s, they built 21 self-propelled tankers sized from 2,000 to 20,000 barrels. The 1930s may have been depression years for many, but certainly not for the Busheys.

Things were going well for the Busheys, so well in fact that by default, in 1936, young Frank got a taste of the good life. Frank's mother and father were to tour Europe, but his mother declined, so Frank went in her place. They travelled in style, out on the QUEEN MARY and returned on the NORMANDIE. A high point was going to the summer Olympic games in Berlin.

The first of Ira's children died on May 31, 1939. It was his youngest son, William. He left his estate of $184,097 to his three brothers and his sister, Ella.

Expansion continued in 1940 when the company bought the old Todd Shipyard on Clinton Street. As with the Mattons, the Busheys were involved in the buildup of the Navy's tug operations just before our involvement in World War II. They sold the CARMELITE, the CONSULTOR, and the COUNSELOR for $181,793 each. Note that all the tug's names start with the letter C. More about this later.

The war years (1940-1945) for the Busheys were typical of what happened to most industries. These years could be described with two words: boom times. As Frank said, "Hell, anybody can make a profit during a war." There was one area that declined rather than expanded. The family reduced their floating equipment to just tugs and oil barges. Gone were the steamers and wood barges. They were now fully committed to hauling petroleum products.

From this point on, all my information came from interviews with Frank Bushey, third and last generation in control of the family business. Frank (christened Francis) was born on October 27, 1919, and lived in the Bayridge section of Brooklyn until 1954. After attending parochial grammar school in Brooklyn, Frank traveled to South Orange, New Jersey, to attend Seton Hall Prep School. According to him, it was quite a trip in those days. Active in sports and a good scholar, he received the General Excellence Medal for four years.

Being the son of a very successful businessman could have affected his personality in making him feel somewhat superior to his classmates. Not

190

so, said Frank. His father saw to it that he didn't develop any ideas of grandeur, and no doubt working summers in the shipyard helped, too.

Frank's decision-making began at the tender age of six. He carried the same name as his father, Francis, and he hated to be called "Junior." Rather than complain or argue for change, he took direct action in a rather unique way. He simply refused to answer to the name Francis. From that time on, if you expected the lad to respond, you had better call him Frank.

Frank graduated from an excellent engineering college, Rensselaer Polytechnic Institute in Troy, New York, in 1941. He went on to one year of graduate work at M.I.T. before joining the Navy.

During World War II, the Armed Services had a rare genius for jamming square pegs in round holes.

Frank Bushey was luckier than many. With his education and background, he was sent to Washington, D.C. in the section of the Bureau of Ships that was responsible for the building and fitting out of shipyards. After 22 months in Washington, Frank wanted out. His choice was a repair ship, but the Navy explained that they would have to train him for that assignment. That seemed a tad silly to Frank, but it was futile to try to figure the Navy's logic. Instead, they sent him to the Brooklyn Navy Yard, which certainly wasn't the worse place they could have sent him.

Typical of the Navy, he was only in Brooklyn four months when they gave him a "permanent" assignment in Hoboken, New Jersey. Frank was about to protest when a wiser head prevailed. His immediate supervisor advised against it. His argument was that protesting too much, you could find yourself in upper Siberia. So Frank went quietly and finished out the war in Hoboken. While he was there, he managed to get in about sixty days sea duty in bits and pieces. The sea duty was one-or two-day shakedown cruises after repairs had been completed on a ship. Frank's Navy career could be summed up as a full one, but with no great war stories to tell his kids. Maybe it's best that way.

Upon returning to the family business, his first job was Repair Superintendent of the shipyard. I'd like to drop back to October 17, 1942, when Frank married Anne Valentine of Harbor View Terrace. I was interested in where and who married them. Frank downplayed it, but I was fascinated. To me it was impressive to be married in St. Patrick's Cathedral in Manhattan by the Bishop of the British West Indies. I had been brought up to believe that only very important people got married there.

When Frank came into the business full time, his father was president and responsible for selling and production. His Uncle Raymond kept the records and when Uncle Ira came in, he took over production from Francis. Since this wasn't a manufacturing concern, the words "selling" and "production" may sound inappropriate but they were Frank's words. They were selling a service and production referred to building and repairing vessels. Uncle Ira died July 29, 1949, and shortly after that, Frank was elected to the Board of Directors and was a vice-president. Frank said the titles were important because it enhanced your credibility with the people you were doing business with. This was one of the reasons Frank became president in 1953 before his father died. The other reason was that his father was 62 and wanted to slow down some.

Uncle Raymond died in May 1956 and Francis died December 18, 1960. The four sons of Ira A. had died in reverse order of birth, as Frank wanted me to take note.

I first met Frank in his office at his Court Street shipyard on October 24, 1978. His office was impressive only from the point of size. It was huge and the decor could be called early warehouse. The furniture was sparse and worn. I recall a large portrait of grandfather Ira on one wall and a display board of old caulking tools from the era of wood boats. The room fit like an old pair of slippers, and I felt comfortable. Well, not completely. Part of Frank's reputation had preceded him, and I was a little intimidated. Solidly built, a ruddy complexion with the map of Ireland spread over it, he reminded me of Tom Feeney.

He was a decisive man who I felt could erupt on very short notice if he were crossed. I was somewhat in awe of him, but in the end, I went away liking him. I did not see the other side that people had spoke about. I did get one hint when I mentioned that some held the opinion that some of the tug operators were out to break Local 333 of the Maritime Union.

I was referring to the 88-day tugboat strike that ended on June 29, 1979. I had repeated talk in the industry that it was Moran and McAllister, not Bushey that was out to break the union. After strongly denying that any of this was true, Frank stated that it was he and Moran who fought the hardest against Joe O'Hare [President of Local 333 at that time], not McAllister.

Frank went on to explain that Mr. O'Hare was difficult to deal with, but then I imagine most union presidents are. Frank told me: "I think Mr. O'Hare neglected to evaluate the capacity of business to meet his demands. In doing that, the Port of New York suffered because our high cost of operation permitted other unions to invade our territory. And in no way was Mr. O'Hare's union strong enough to say to the loading docks,

Frank Bushey.

Bushey Shipyard, Brooklyn, New York.

Example of a double tow in the Hudson River.

Typical Bushey-built tug with two stacks. One for intake, one for exhaust. Photo by George Michon.

'You better not load that,' so he penalized the operators in the Port of New York."

Our next subject was stockholders and dividends. As with the Feeneys of Kingston, all the Bushey stockholders were family. There was one very significant difference between Ira Bushey and Thomas Feeney in the distribution of stock. All the Feeney stock was distributed evenly to all his children. Not so with Ira. The largest share went to Frank's father, Francis. I asked why. Frank replied, "My grandfather appreciated my father's abilities. Maybe he was smarter than the rest of us." This remark explains why Frank and his father became presidents. Frank gave no hints about any resentment that might have been harbored by Frank's uncles.

When the stockholder's meeting was held in the shipyard office, about 10-15 attended. These meetings were no big thing until after his Uncle Raymond died. By then, the next generation was old enough to be interested in them. The degree of interest in stockholder's meetings increased in direct proportion to the rise in profits. They paid a fixed dividend each year and were reluctant to change it. Frank said, "In a family business, it's almost impossible to cut a dividend. If we increased it and had to cut back, it caused trouble."

I shifted to a question of why they maintained an office at 500 Fifth Avenue, which served as the office for Spentonbush, the operating arm of Bushey. Business was conducted in three other locations as well – two lunch clubs in Manhattan and one in Brooklyn. I wondered aloud why he didn't just use a restaurant. Frank quickly straightened me out on that point. "The type of restaurant that you would conduct business in is too expensive. You have to buy your way into a decent table."

Frank followed the pattern of most other executives by belonging to and being an officer in most of the maritime transportation organizations. When asked if he found any part of business boring, Frank replied, "No business is boring if you are in a position to make decisions." A candid statement I did not expect to hear, but I was pleased with his answer.

I was interested in finding what, if any, advertising they did. Frank pointed out that putting an ad in the *New York Times* would have no value. Bushey does what Frank called defensive advertising by placing ads in trade journals. The purpose is not so much to drum up business, but to let the industry know that Ira S. Bushey and Sons was alive and well.

We moved on to the division of their fleet operations. Starting in the 1940s to the 1960s, the business was evenly divided between canal work in the summer and coast work in the winter, utilizing the same crews and equipment. After the opening of the St. Lawrence Seaway in 1959, the predictable happened, a steady slide to more and more coast work. Frank's reaction to questions about the canal itself matched Tom Feeney's responses – negative. "The canal is obsolete, no improvement in years, and the railroads are a factor in its disuse." He pointed out a Catch-22 situation. "The state said, 'show us the traffic and we'll improve it.' The shippers say, 'it's so obsolete we can't make a profit using it.'"

Frank echoed Tom Feeney's words about limited tonnage compared to the Western River's system. Also, like Tom, he tried for years to get the Federal Government to take over the canal, but Governor Nelson Rockefeller was against it, and they couldn't get by him.

Before going on to other things, I asked Frank for a comment about his father, something people might remember him for. His response was slow in coming. He wasn't able to pinpoint any one thing. This happened often to me when asking for someone to speak off the top of his head on this subject. Finally he said, "Excellent witness." His father was in court often as an expert witness in defending tug and barge owners against the insurance companies. He would explain that a particular accident could have or not have happened in the manner presented.

The last of over 600 vessels built was launched in 1966. It was the tug BOSTON and had a special feature, an automated engine room. The engines were started, run, and stopped from the wheelhouse. Instead of an engineer for each of three watches, the BOSTON carried one. She is one of two tugs used as a model for union negotiations concerning automation. The reason for not building anything after 1966 was unions. The labor, became excessive, according to Frank, because the unions forbid their members to cross trade lines. As an example, an electrician was not allowed to paint. Neither could a welder be a pipe fitter. The advantage for the shipyard was flexibility in scheduling work. The union won the battle and lost the war and their jobs.

During the life of the Bushey organization, they bought out Red Star, Williams Blue Line, and Sheridan of Philadelphia. The Sheridan takeover was a bit different than the others. Bushey first became involved with them when they built a tug for Sheridan and shared ownership. The ties became stronger until finally Bushey bought them out in 1972.

Jumping around a bit, I asked why there weren't more pushboats such as the MOHAWK and the ROCKLAND COUNTY in our area. It's almost 100% pushboats on the Western rivers. Pushboats are fine for the canal and river, but not seaworthy enough for Great Lakes or coast work. The conventional tug can work all these areas. More about the MOHAWK in the chapter on special vessels.

My next area of interest really struck a nerve in Frank, and has apparently been a mean cross for him to bear. What could bring out such a volatile reaction? It was the mention of Government Regulatory Agencies such as the EPA (Environmental Protection Agency), OSHA (Occupational Safety and Health Agency) and even the U.S. Coast Guard. He warmed up by saying that we don't need OSHA because New York State has plenty of laws and regulations to do the job. He had some unkind things to say about the Coast Guard, but he wasn't too specific. Frank hit full stride when talking about the EPA. He feels that the EPA was the worst thing to happen to American industry. Production has gone down and cost risen because of the billions of dollars spent on complying with their regulations.

Frank gave an example of how the EPA could improve. They should agree on a design for scrubbers to reduce smokestack emissions, then say to industry, install this unit and we will bless your operation. This is not the way it's done. They say install it, then we'll take some readings. If the readings aren't OK, you'll have to install something else. Quote Frank, "Who the hell is going to invest that kind of money with no guarantees?"

Frank continues, "Maybe we don't need snail darters (a tiny fish that caused a dam project to be stopped in Tennessee) and maybe we don't need some kind of eagle anymore. I'd like to have them around, but when they're gone, they're gone."

"Big industry didn't kill the dinosaur. We had little guys with stones." I was delighted with his last remark on the subject. "If God intended that we never change the environment on this earth, we'd sure have a hell of a time with dinosaurs running around Brooklyn."

Frank's last shot was at the Internal Revenue Service. He said they look at a family-run business with a microscope. The IRS assumes that you have to be cheating in some way. This response was triggered by a question about his salary. Frank felt that the same job in a public held corporation would pay more, in spite of what most people think. He could be right because other owners conveyed the idea that they could not pay themselves large salaries.

To my questions about today's workers vs. the ones in the 1940s and 1950s, Frank was quick to say that today's workers don't work as much, know as much. or care as much. His answer was predictable in that I have found each generation thinks the preceding one was a paragon of virtues. The thought must have struck Frank the same way as soon as the words were out. He chuckled when he remembered his father saying the same things to him.

I did have his complaint about the lack of loyalty of today's workers reinforced a few weeks later. It occurred when I went to interview an old time employee of the Bushey's. It was Bert Connor and before I could even turn on my tape recorder, he let me know where his loyalties lie, "You won't get me to say anything bad about the Busheys," he said right away. I get very suspicious when anyone starts a conversation in that manner. It sent a signal that I wouldn't get much information from Bert. That's exactly how it turned out. He was noncommittal and more evasive than a politician. He did drop his guard once. I asked if Frank Bushey was a tough boss. "No tougher than me." He went on to explain that he could be demanding when the need arose. What Bert didn't understand when I asked for information about the Busheys was that I wasn't looking for skeletons in the closet.

Moving on, I learned that Frank hadn't taken a ride on one of his vessels since he was 14. Not enough time, but I would say lack of interest was the real reason. In this respect, he was no different than the Feeneys or Mattons. He did love boating, but a different kind, motor yachts. The family owned a 58-foot Elco before World War II. It was donated to the Navy to use during the war. Frank bought his first cabin cruiser, a 38-foot Chris Craft in 1965. He moved up to the Cadillac range with his purchase of a 43-foot Hatteras yacht in 1980.

The only family tradition Frank could think of was drinking champagne on Christmas. To keep it respectable, they didn't start drinking until 11:30 a.m.

The one regret in life that Frank was willing to talk about was a sad commentary of the times and another jab at the state and federal governments. Inheritance taxes were so high that it didn't make any sense to continue in business. So on June 24, 1977, another part of New York State Maritime industry became a part of history. After 77 years, all was sold to one of our largest oil companies – Amerada Hess. Why would Hess buy into a dying business? Speculation is that Hess wanted the oil terminals, and that does make sense.

The most visible change came in the color of the boats – from a dreary dull grey to Hess "green" which is no great improvement. The Bushey grey started during World War II, and Frank never changed it. I feel they carried the inconspicuous to the ridiculous. Besides the dull paint, they didn't put the company name or initials on their tugs. This offers a clue to Frank's personality, plus the ghost of his grandfather. As Frank said, "If we are going to go against my grandfather's philosophy, we don't need to flaunt it."

The trademark of unique bright color schemes of the Moran, Turecamo, and Mobil boats was not for the Busheys. I keep falling into the trap that owning and operating tugs and barges is romantic and glamorous. To Frank Bushey and his ancestors, the only thing glamorous was the bottom line on the yearly balance sheet. As his father once told him, don't be afraid of taking a profit.

The rest of the Bushey story is a list of firsts. They are firsts, however, with this disclaimer: to the best of Frank Bushey's recollection. All the firsts mentioned in this book carry this same disclaimer because records are scarce and memories are not the most reliable of sources. Anyway, here are just a few notable Bushey "firsts."

1929

The 4,000 barrel self-propelled tanker, IRA S. BUSHEY, was the first all electric welded vessel given a coast-wide license.

May 24, 1939

The fifteenth steel tug since 1935 was launched today. The CHAPLAIN was a first for the Busheys in that the engine room was soundproofed. Francis Bushey got the idea from a visit to a radio station. Quote Francis, "Steel ships are like drums. The noise is terrific and affects the efficiency of the crew."

1964

Busheys initiated the first "Jumbo Barges" to be used on the canal. They took two tank barges, HYGRADE #14 and HYGRADE #8 and lengthened them to 293' 9". They were equipped with bow thruster steering and propulsion units, making the overall length 297' 9". This left only 2' 3" clearance in the locks. Even with the added time for double locking, Frank thought it was worth it for the extra capacity gained.

1966

Who was the first to push two oil barges in the canal? This is the most controversial first, since Moran, Kehoe, or Coyne could claim the honor. It was really an extension of the jumbo barge idea. The problem wasn't in the double locking but at the terminal where the cargoes were to be unloaded. For example, you could not drop one barge at Burlington, Vermont to go up to Plattsburg, New York with the remaining barge. The terminal at Plattsburg wanted the tug to standby in case of foul weather. Another reason for ending this operation both on the Northern and Western Branch was not enough volume to make it profitable. For this trial, they used the tug CHEMUNG and barges HYGRADE #2 and BLUE LINE #108.

We are left with a couple of unanswered questions. Why did the Busheys name their tugs in the fashion they did. Frank could not tell me. They had a fascination with the letter "C," religious and Indian names. Some examples:

CHIPPEWA	CHEYENNE	CHOCTOW	CHEROKEE	CELTIC
COMAMCHE	CREE	CROW	CHANCELLOR	COMMODORE
CARMELITE	CARDINAL	CHAPLAIN	CANDARGO	CORTLAND

For size and staying power, no other outfit came close to the record compiled by the Busheys in their barge canal operations.

There is really no mystery to the family success. They followed an age-old business rule, adapt or die. To get a real jump on one's competition you must recognize the changes in your industry almost before they happen. When the Busheys did see the changes coming, they didn't test the waters with their toes, they dove in head first. There is another reason for their success, too, for they were able to use their fleet the year round, not just the seven-month canal season, which was the downfall of many of their competitors.

CHAPTER 12

THE FEENEYS OF KINGSTON

Important as the Feeneys are to the history of the Barge Canal, the frustration of trying for four years to make contact with them was enough for me to seriously consider throwing in the towel; but I felt it important to include their story. Bernie and Tom Feeney, sons of the founder of the Kingston based business, ran Feeney enterprises. Despairing after several unsuccessful attempts to meet with Bernie, who presumably had the most knowledge of the company, I finally got lucky. From Bernie's son, Ed Feeney, a lawyer in Kingston, I was put in touch with Tom, an equal partner with Ed's father and, as it turned out, equally versed in the company business. I learned, by the way, that Bernie was a compulsive record keeper and kept some notes that just about raised trivia to an art form. Yet there was a positive side to some of his infatuation with minutiae and records, for he also had sketches of the construction of the wood barges his family had built – invaluable historical information.

Following my first contact with Ed Feeney, I set a date to go to Kingston and talk with Tom Feeney, his three sisters, and Bernie's son as well. Within a week I was in this Hudson River town.

Tom met me at his door and he looked as I imagined he would. A big burly man, a little past his prime, but it didn't take much imagination to see that he must have been someone to reckon with in his earlier years. His firm handshake and the look in his eyes said welcome. I usually shy away from making instant character judgements, but I did this time. By the time I reached his kitchen and met his lovely wife, I felt like a member of the family.

I started the conversation with a compliment. It's an instant ice breaker if it comes from someone the person respects, and it did. The compliment came from Captain Jack Maloney, who had pushed Feeney barges for Matton. He had many nice things to say about Tom, but most of all he remembered the times when Tom took him out to dinner in Buffalo. This was when Tom was a "runner" for the business. Another compliment came from Cliff Arnold, a mate now with Sears Oil, who said, "Feeney bargemen were the best and probably Connors the worst. Feeneys could lock through slicker than anyone because they all worked as a team."

With this as openers, I started to ask about the beginning of Feeney Enterprises. The date was either 1867 or 1868 – no one is quite sure which – when Owen and Elizabeth Feeney and their children arrived in the United States with the hope that a better life could begin. The potato famine had driven many families like the Feeneys from Ireland. They settled in the village of Alligerville, New York, about ten miles south of Kingston.

Owen,a stonemason, found work at his trade on the Delaware and Hudson Canal. His oldest son, Bryan, was hired as a waterboy. The now defunct D & H Canal ran from Kingston to Port Jervis on the New York border then up to Honesdale, Pennsylvania. The principal product carried was coal. After completion of the canal, Owen moved his family to Kingston, where he helped build City Hall and St. Mary's Church. Bryan bought a team and took to the tow path of the canal. A few years later, he met and married Bridget Knox, and from this union came 11 children. As his sons reached the age of ten or so, they were hired out to drive teams on the canal. One of those boys, named Thomas A. born on January 28, 1872, went on to be the founder of the Feeney boat business.

Following the example of his peers and the pressing necessity of just surviving, Bryan could not afford the luxury of letting Thomas spend much time in school. Thomas did go to school in the winter for a few years, but he never learned to read or write, except to sign his name. The fact that he was illiterate never became the curse it was for Frank Coyne. His daughter Rose said his Dad believed that whatever his lot in life was, he accepted it as God's will.

Sometime after reaching manhood, Thomas gave up the tow path to learn the trade of building boats. Leaving the tow path also improved his social life. His sisters worked in a factory in Kingston and had a friend that they thought might be just right for Thomas. Her name was Rose Woods, and they were correct She and Thomas were married on December 20, 1899. In 1901, their first child was born, Philip. Bernie was born in 1902, followed by Margaret in 1904, Marion in 1908, Rose in 1910, and Tom in 1914.

These were very lean years for the family because boat building was seasonal. Shortly after Thanksgiving, they shut down. Now they all prayed for an early freeze. With the boatyard closed for the season, the only work available to Thomas was for a local ice company. As soon as the river was frozen over, Thomas got up each morning at four o'clock, walked three to four miles to cut and pack ice for ten hours a day. This seasonal job changing went on even after time and experience had won Thomas a promotion to foreman at two different boatyards. His whole world changed in 1917.

Thomas was approached by William Rafferty, a Kingston businessman, with the proposal of starting a boat building business. It must have been a scary idea for Thomas. He was 47 years old, with a wife and six children to care for. Building a canal boat was one thing, but to run a business would be a quantum leap for Mr. Feeney.

The conversations that he and Rose had would be duplicated some years later by Frank and Teresa Coyne. Both families were in similar circumstances. Both men were illiterate, worked with their hands, and had scant knowledge of how to run a business. Both had large families and would have to give up steady jobs. In the final analysis, I believe the deciding factor was the faith and confidence shown by their wives. Both were willing to make whatever sacrifices necessary for their husbands to start a business.

The decision was made to accept Rafferty's offer, and Thomas's brother-in-law, Philip Woods, became a third partner. A key element in this partnership was Rose Feeney's contribution. Since their marriage, she had saved approximately $1,300! With Thomas earning about $3.00 a day, it was a small miracle how she did it.

The firm was named W. F. & R. Boatbuilding Inc., located at East Strand and Tompkins Street, and had a crew of about 70 men, they built wood barges, scows, coal boats, and derrick barges. In 1929, they moved to their present location at 613 Abeel Street.

The Feeney girls shared a chore in those early years. Each day at noon one of them would rush home,grab a lunch bucket with hot food for their Dad, and run to the boatyard. Cold sandwiches were definitely not the order of the day. All the boys worked in the yard during the summer vacations. There was no special treatment for the Feeney lads; in fact, Tom thought his Dad gave him and his brothers the hardest jobs. Hard physical labor was a virtue to be proud of as far as Thomas Feeney was concerned.

The first annual board meeting was held on July 2, 1918. Thomas was elected President with a salary of $1,500, Philip was Vice-President and Secretary at $1,500, and the "money man" (Feeney definition) William Rafferty, Treasurer at $2,000 a year. In 1920, Woods and Feeney's salaries were doubled, but Rafferty's was tripled!

By 1920, Bernie and Doc (Philip) were working full time in the yard. No one could tell me how the nickname "Doc" started, but it was all Philip went by during his short life. Doc started in the yard while Bernie went into the office. I never got to ask Bernie himself, but the family felt that he was not completely happy in the office, yet as a dutiful son, he accepted his Dad's decision.

Business was good from the start. They built an average of 14 barges or scows per year from 1917 through 1931. The year 1923 was a super one with 23 built. They had some stiff competition during these years from Matton in Cohoes and Hildterbrandt, Schroonmaker, Dwyer, and Lenahan in Kingston.

Also during these years, one member of the firm was not physically in Kingston. Philip Woods had a coal mining business that he operated from New York City. 'When he became a partner of Feeney and Rafferty, the company paid half his office rent. He was their "salesman" as the Feeneys referred to him. His job was to drum up business. In addition to his efforts, Thomas Feeney, sometimes with either Bernie or Doc in tow, took the 6:30 a.m. train to New York every Wednesday morning looking for work. Any contracts secured were finalized with just a handshake. This was the accepted practice in those early years. The fact that he could not read or write had no bearing on this practice. Ralph Matton, with a college education, did the same thing.

Doc was the yard boss under his Dad's watchful eye. Bernie started as a timekeeper and when the bookkeeper left, he moved into the office full time. He had two excellent teachers in Mr. Rafferty and the accountant, W. F. Davis. Being a timekeeper required Bernie to keep accurate records, and this may be where he got his passion for keeping records in minute detail. Another part came from his Dad.

The business grew and prospered, but there were problems brewing. The partnership ended in a bitter lawsuit with Philip Woods that was resolved in Mr. Feeney's favor in 1934. Also, Mr. Rafferty died before the suit was settled, and his widow was anxious for Mr. Feeney to buy her stock, which he agreed to, though it took a number of years to accomplish. I was curious about the nature of the suit, but I did not pursue it. Although I could have gotten the records from the court house, I decided to respect the family's wishes that I not go into the details. I don't regret that decision. After the split up, Reliance Marine was formed. This was the first of many corporations that would become Feeney Enterprises. The last Feeney son entered the business in 1932. Young Tom started as a runner. More of that later. 1933 may have been a year of a first for Feeney. They built a 115 ft. × 36 ft. wood scow for Connor's Marine. It was standard except for one feature – steel tanks were installed on the deck to carry molasses. Connors became one of the largest carriers of this product. Incidentally, it is now the last food product being hauled on the canal.

There was little change in the construction of wood boats over the years.

The Feeney family. Left to right: Thomas J., Rita A. (wife), sisters Margaret, Marion and Rose M., Bernie and wife Florence.

Thomas A. Feeney.

Rose Feeney.

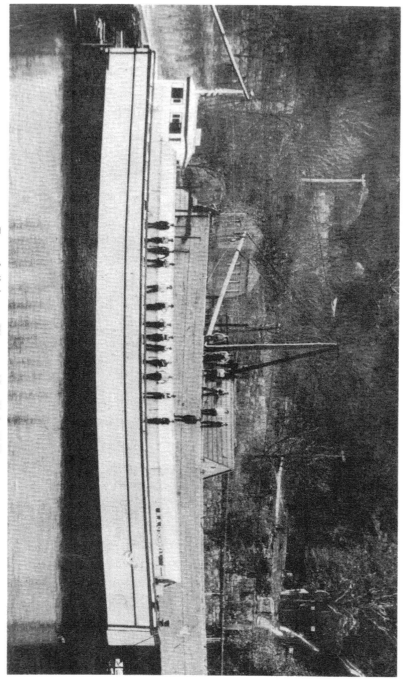

Feeney-built barge, Kingston, New York, 1923.

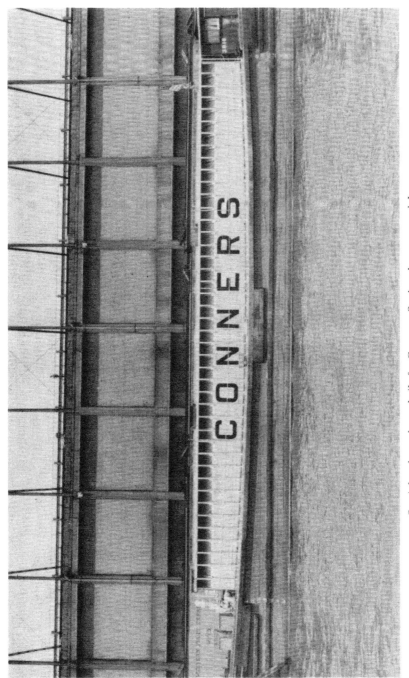

Special molasses barge built for Conners. Steel tanks on a wood barge.

Feeney Shipyard, Rondout Creek, Kingston, New York.

They were not built using blueprints as we would today. When a change was desired, the Feeneys would explain to their highly-skilled mechanics and they were able to translate the idea into reality without detailed sketches or drawings.

In addition to building boats for others, they built and bought for themselves. Starting with three barges in 1922, they built the fleet up to where they were running an average of 26 barges per year on the canal and Hudson River. Most of this phase of the business was over by 1965. I was curious as to where they kept all those boats around their boatyard on tiny Roundout Creek. Well, they didn't. The fleet was tied up at the grain elevators in Hoboken, New Jersey, and at the foot of Columbia Street, Brooklyn, New York. For many of those years, Connors towed the Feeney fleet on the canal, and the famous Cornell Steamboat Co. of Kingston towed their boats on the Hudson River. They maintained a close working and personal relationship with Harry and Arthur Connors throughout their lives. So much so, that Bernie Feeney went with Arthur Connors to pick out Arthur's burial plot.

By the early 1930's, Thomas Feeney was getting concerned about the cost of towing. During the depression years, there was a government agency known as the Reconstruction Finance Corporation. One could submit a program for improving his business and apply for funds. They would contact his local bank and if both approved, he would get a loan backed by the U.S. Government. Since Thomas Feeney's credit was excellent, he had no problem getting a loan. He bought the hull of SOCONY 6 for $5,500 and rebuilt it and installed a diesel engine. They named the tug THOMAS A. FEENEY and in 1936 started towing some of their own barges. FEENEY GIRLS and ROSE A. FEENEY were brought out in 1944. The next two were PHILIP T. FEENEY in 1950 and MARGARET FEENEY in 1951. With the exception of the MARGARET FEENEY, these four tugs were built in the late 1800's. All had small diesel engines installed in them, none higher than 550 hp. The Feeneys spent a lot of time and money rebuilding these tugs.

At this point, I should explain what a runner was. A runner was a key person with companies that ran large fleets of tugs and barges. His primary function was to keep them moving. Delays could be caused by high water, leaky boats and drunk crewmen. Today the problems aren't the same because you have one steel tug pushing one steel barge. Today's tugs have a more powerful engine to better handle high water with its faster currents, and steel barges don't damage as easily as the wood ones did.

Also in the old days, there were more men to deal with. With a steam tug and six barges, there were 21 men. Today, there are eight if the tug has an automated engine room (i.e. MORANIA 6), or nine if it is not. Two of these men ride the barge.

A fleet consisted of four to six barges. If a company like Feeney had more than one fleet, the runner became just that. He was constantly on the run. So between doing repair work, coaxing the tug captains to move, and dragging bargemen out of the saloons, it was not an easy job. As Tom Feeney said, "You were half caulker and half corker." Although his brother Doc was the first runner, Tom learned the work from his Dad. Tom admitted to being a little wild as a youth, so it was five years before his Dad left him on his own.

Jack Maloney told me a story about Tom that bears repeating. Once when Jack's mate was on a train to Tonawanda to catch the tug, he struck up a conversation with the man sitting next to him, he told the man all about his job and that he might sink one of the barges. The man turned to him and growled, "You do and you'll get one in the mouth." Until that moment, he had no idea he was spouting off to Tom Feeney. When the mate related the story to Jack, he got another shock. Jack said that Tom Feeney was dead serious. It turned out to be a self-fulfilling prophecy. Entering the first lock, the mate rammed the head barge into the lock wall. He was terrified that Tom would follow through with his promise. Tom didn't, of course, but voiced his displeasure with some vigor.

During these years, Feeney boats carried grain and pig iron east, newsprint from Canada south, scrap iron sulphur, oil, and waste paper west. Using newsprint as an example will give us some idea how these contracts were born. The Feeneys learned that the *New York Daily News* was moving an increased amount of newsprint from Canada to store in Brooklyn. The Feeney's were completing the construction of two steel barges and went looking for the business. With some adjustments to hatches, they got the contract. They had some competition in their own backyard. Schroonmaker in Kingston had two motorized units hauling also. The contract lasted from 1945 to 1950. The end came because the stevedores in Canada and New York were feuding over the loading and unloading. The *Daily News* resolved this by buying their own self loading/unloading ship. End of contract for Feeney.

An era ended in 1940. They built their last wood boat and started on the first steel barge. It would be two and a half years before they launched her. The war made it very difficult to obtain steel. Oddly enough, the first steel barge was also the largest one they ever built. It was the JOE HERR-MANN. Built for Connor's Marine, she was 225 ft. × 42 ft. Joe Herrmann was a partner and the right hand of the Connors family. Tom Feeney admired Joe and claimed it was he that advised Connors to sell out just before the canal business took a nose dive due to the opening of the St. Lawrence Seaway. The JOE HERRMANN may have been a first in that she was registered as a combination barge. She could carry a liquid cargo such as molasses or switch to a dry cargo such as wheat. Forty years later, she was still in service.

An end to an era triggered an obvious question. Did the Feeney boys differ with their father on how things should be done and what was the outcome if they did? Of course they did, and the building of the JOE HERRMANN is a good example. After it was launched, the boys wanted to build all barges of steel. Their father was a wooden boat man and was not in favor of it. After several discussions, he agreed to build two steel barges to see how things developed. He was pleased when they were launched end – of discussions.

A prelude of tragedy began in 1938 when Doc took sick. Doc was his Dad's right hand and Tom felt Doc was sharper than either he or Bernie in overall knowledge of the business. Doc was only 40 when he died in 1941. It was shortly after Doc's death that their father elected to stay home and look after his ailing wife. I view that as an act of love and devotion, yet knowing the Feeneys it comes as no surprise. He could have hired a full-time nurse to look after his wife, but he chose to fill that role.

Just because he stayed home does not mean he gave up the reins to the business. Up to three months before he died, Bernie would report to him morning and night to discuss business. If there was a problem that required his personal attention, he would go to the yard, and he and the rest of the family never missed a launching.

Thomas A. was a man of simple tastes and pleasures, all of which revolved around his family. His faith was the rock that sustained him throughout his life. The same could be said of the entire family. As an example; all during Lent, everyone gathered in the living room after supper, knelt and said the Rosary. It didn't matter who was there, kith or kin as Tom put it. All the Feeneys I talked with made a point of the rich religious upbringing they had. They were not Sunday's only Catholics. Tom volunteered that he believed that prayer and faith were the reason for whatever success they had. Who's to say him nay, with almost all of their competitors long gone from the scene. The only surprise for me was that in an

Irish Catholic family of such religious convictions, not one became a priest or nun.

I received another surprise when I asked what they built for the government during World War II. The answer was nothing! Too much red tape, they decided, yet war contracts were what gave others in the industry such as Bushey and Matton a much needed shot in the arm. Instead, Feeney had a large contract with American Literage Corp. for repair work.

By the end of World War II, the Feeneys were running two tugs and 26 barges on the canal. In addition, the construction and repair of barges and scows continued at a steady pace and does to this day. Although the number of tugs grew to five by 1961, the barge numbers ranged between 24 and 30 until the mid 1960s. By then, the canal business had dropped dramatically. The opening of the St. Lawrence Seaway took its toll on Feeney as it did all the others.

A few years prior, Bernie and Tom saw the handwriting on the wall and started putting their scows out on long leases. These scows were not canal boats. These steel scows were hauling mostly aggregate on the Hudson River, in Long Island Sound, and along the coast. They were not affected by the opening of the St. Lawrence Seaway.

I asked Tom if he would elaborate on this decision. "The Future of the canal didn't look that bright – no help from Albany and I wanted the Federal Government to take over the canal. It's a play thing for the state politicians. The most one tug can push on the canal is 2,500 tons, while on the Mississippi they push 50 to 60,000 tons. We felt, and rightly so, that scows would always work, so we invested our money in them and they are still out on long lease."

Tom added other reasons to get out of the towing business: the aggravation of labor contracts and the tremendous rise in insurance premiums. During the heyday of their towing business, they bought out two defunct companies. They were bought for the only assets they had of value – their ICC rights. These licenses granted by the Interstate Commerce Commission are required if anyone operates a barge, tug, truck, etc. outside of the boundaries of their state. The Feeneys bought the rights from the Empire State Corporation in 1952 and Murrary Transportation in 1953.

During my research for this book, I was fascinated by the relationships between people like the Feeneys, Matton, and Bushey. They were competitors, customers of each other, and friends to boot. Many of Feeney's boats were repaired on Matton's drydock, Matton towed Feeney barges. A fierce competitor yet Bushey helped the Feeneys out when they were in a tight spot. When Feeney bought their first tug, THOMAS FEENEY, they ordered a Fairbanks-Morse diesel engine for approximately $25,000. This was in the height of the depression, 1936, and they had never done business with Fairbanks before. Even though they had a good credit rating, Fairbanks kept stalling them. Bernie and Tom talked with the Busheys who had just ordered five engines from Fairbanks. They in turn phoned the Fairbanks factory and told them if they did not honor the Feeney contract, they would cancel their contract as of that minute. Needless to say, the Feeneys got their engine.

Throughout the years of canal operations, the bulk of Feeney cargos were the usual items – grain, scrap iron, soda ash, newsprint, and pig iron. I'd like my young readers to know that there were other things carried that were a little out of the ordinary such as Army tanks, missiles, bridges, lock gates, burlap, coke (not the soft drink), chocolate bars, crude rubber, and fluorspar for making glass.

The good feeling about the growth of the business in the 1950s was marred by the death of the founder's wife, Rose, in 1955. Thomas A. himself died in 1959. He left the business in excellent shape; all the children shared equally in the estate. In 1963, Tom and Bernie bought out their three sisters. Their thinking, according to Tom, was that it would be better for them to have cash in hand to invest as they saw fit.; This way if the business took a nose dive, they would be protected and this was what their father wanted. The boys were always to look out for their sisters.

The Feeneys lacked one thing that would make them a complete marine organization – a drydock, according to Michael Sweeney, a grandson of Thomas A., in his history of Reliance Marine. It was not a new idea, but the higher costs of drydocking in the New York City/New Jersey area caused Bernie and Tom to take a very serious look at the cost of building one. The other consideration was the cost of the repair for the boat once it was on the drydock. They were farming out work they were perfectly capable of doing themselves.

After much discussion about types and costs, they traveled to New Orleans to see one that operated on the principle of a submarine. Instead of using a water pump to empty the drydock, air pressure was used to force the water out. This method is much faster and gasoline or diesel driven air compressors are cheaper to operate than electric-driven water pumps. There is an extra bonus because the air compressors can be used for sandblasting the scows prior to painting.

Tom told me that small outfits like Feeney saved nickels and dimes and that helped them survive. An example is their method of dredging and building the drydock. They did their own dredging and dumped the spoils on shore. Using a small front end loader and an old dump truck, they hauled the material away. They wouldn't be able to do that today. An environmental impact study would have to be done. Many government agencies would be involved and it could take up to three years before they go ahead if at all. The drydock itself was built by welding two used barges together with some modifications.

Talk in this vein led me to ask Tom what were the differences that caused the most headaches between the periods of 1930 and 1950 vs. 1960 to 1979. He came back quickly and forcibly. "Two much bureaucracy – New York State compensation insurance – bunch of thieves. Plus the paperwork caused by those nuts in Washington. The problem isn't in securing business."

I asked for comparisons on construction between the old days of wood and now. It took an average of forty days to build a wood barge and they built two or three at a time. Today, with some exceptions, steel scows take three months to build. Twenty men use 300 tons of steel to do this. The scow will sell for about $300,000 with a net profit to the Feeneys of approximately $10,000. Wood barges;sold for about $10,000 and I'll estimate a profit of $400 to $500.

The Feeneys have always paid a little above the average for the area. Thomas A. Feeney and John E. Matton had the same philosophy about work with one exception. Both demanded a day's work for a day's pay and no shirking, but Thomas A.'s pay scale was not rigid. Sometimes he would direct that 25 or 50 be added to someone's pay envelope, not a practice John E. would approve of. 25 or 50 doesn't sound like a great sum, but when the going rate was $3.00 for a ten-hour day, it was significant.

I had just a few more questions as my interview with Tom wound down. What did he like most and least about the business?

Least – "Bickering within the family." I'd heard the same thing from Frank Bushey.

Most – "I get intense enjoyment out of having a good day, a sense of accomplishment."

On a scale of 1 to 10, Tom rated money as a 9. It wasn't important as long as he had enough to eat. Tom said his only regret other than personal ones was that they should have built the drydock 20 years sooner. His voice trailed off mumbling something about hindsight.

Thomas J. Feeney, son of the founder, touched me with his parting

218

words. "If only Pa could come back for a couple of hours and see what we've done with what he left us. Remember the whole theme of this thing started with my father. We deserve some credit for carrying the business on, but let me tell you, the blood and guts were his." Leaving Tom, I headed for 12 Wilkie Avenue, the home of the Feeney girls.

I spent a delightful afternoon with Rose, Marion, and Margaret. All had professional careers and chose not to marry. Rose was a nurse, her sisters were school teachers, and they stayed in the family home with their parents until their parents died. They still all live together in a lovely home on Wilkie Avenue.

Although Rose is the youngest, it became apparent early on that she would be the spokesperson for her sisters. They were all eager to talk about their parents. In fact, sometimes all three would talk at once. As you may have surmised already, the family home was the center of all activity for the Feeneys.

As an example, all the grandchildren would come to the family home each Sunday after mass for breakfast. They followed a custom of the men eating first and then the women sat down. This may have been an old Irish custom, but it was new to me. Most important occasions were held at "Nan and Pop's" house. Thomas A. and Rose's world revolved around the family, so much so that they never took what might be thought to be a normal vacation. A couple of days once in a while to visit relatives in Oneonta, New York. One day of the year was extra special, their mother's birthday. She loved a big fuss, so everyone went all out to honor her on her day. She was made to feel and was, in fact, a queen.

In many families, visits to the grandparents tapers off dramatically by the time the children reach high school. Not so with this family. Ed Feeney, son of Bernie, told me of coming home for the weekends from Albany Law School and making his first stop, not at home, but to see his grandparents. Some cynics would say these were duty calls. I don't believe that for a second. Anyone talking to Ed Feeney could see how deeply he loved and respected his grandparents.

The only Feeney daughter to be involved with the business was Rose. In her last year of high school before entering nursing training, she worked in the office. Since her retirement, she has gone back to help a couple days a week in the office. Rose would become my contact to Bernie since he steadfastly refused to be interviewed by anyone.

There was one part of the business all the Feeney women were involved in, boat launchings. They seldom missed one. I got the distinct feeling that they were as thrilled to participate in the launching of a scow as most people would be to see an ocean liner or battleship launched. I think that says more about the family than all the words I might put down. One last word on launchings. Thomas A. saw and directed hundreds of them, yet at 75 with a broken ankle, he was taken in a wheelchair to witness yet one more scow slide into Roundout Creek.

From Rose, I have Bernie's memories of his mother. "When I talked about our mother, he laughed and responded, "She was the Major-Domo." The reason for this comment is as children we were made aware that my mother was in charge of the household and all requests and discipline were under her supervision. Our father was firm in the fact that he would not make a comment on something that occurred when he was working. Bernie's last comment was, "You know Mom had to have been a very good manager when you think of all the people we had in and out of our house and she provided and took care of everything."

On Pop, Bernie said his first recollection was working as a water boy during the summers and on Saturdays when going to high school. He told me, "Pop was a perfectionist. As long as you were told and did it right, you were O.K." He said Pop knew every inch of a boat;it's cost, how much production should be completed daily, never ordered beyond his needs, and demanded that every tool be in its place at the end of the day.

I asked what happened if you didn't do something right. He said he would show you again, but you better not make the same mistake twice. He said it was wonderful working with a man who loved his work, knew every detail and did it so well. Bernie really valued the training he received. As Rose pointed out, you could now see where Bernie acquired his need for detail. He never ventured into anything until he knew every aspect of it, whether it was business, board member, or whatever. As we have seen, not only were the records kept in detail, but never thrown away.

Rose asked Bernie what he was most proud of. I expected an answer connected to either business or his work in the community. He was the most proud of securing his high school diploma. Rose explained that Bernie left high school in his senior year despite Pop and Mom's request that he return. Pop then put him to work in the boat yard. The summer before the next school year started, the principal and one of his teachers got in touch with Pop and Bernie and convinced him to return. When he graduated, he was awarded a $5.00 gold piece for mechanical drawing. The high point in his life, he felt, was his appointment to the board of the Kingston Trust Co. (now Key Bank). He was only the second Catholic to be selected

Feeney barge, MIMI, unloading newsprint, New York City.

Deck loaded with newsprint, New York City.

The perils of owning a wood tug.

223

224

and three years later, he was asked to join the Executive Committee.

Bernie was deeply involved in civic matters as the following list will show:

1. Member of the New York State Waterways Association, served on many committees.
2. Member of the New York Tow Boat and Harbor Carriers Association and served on the Labor Relations Committee.
3. Appointed by the Governor to the Ulster County Community College Board.
4. Board member for 25 years and President for five years on the Kingston City School Board.
5. Life member of the Kingston Kiwanis Club.

In addition, he belongs to the Kingston Knights of Columbus, and local yacht and country clubs.

Since the opening of the barge canal in 1917, more than 100 corporations have come and gone. Only two remain with members of the original family still in control of the business. One is the giant Moran of New York City, and Feeney of Kingston. At the time of my interview with Tom Feeney, I didn't see any evidence of a third generation ready to take over. He assured me the family was working on the problem. There is an ironic twist to the story of the continuity of Feeney Enterprises.

Tom had told me that Bernie has always discouraged Tom's children from entering the business. Talking with Bernie's son Ed in 1979, he said decisions would have to be made in the next couple of years because of his father's age and his Uncle Tom's health. Decision time came with the death of Tom Feeney. The business was liquidated on November 27, 1982. The new operators are Tom's sons, Thomas R. and Brian. I was very happy to hear that. That may sound odd coming from someone who has no vested interest in who runs Feeney Enterprises. I think we all should be interested from this viewpoint. I can't articulate it, but I feel it's important that Feeney hasn't been swallowed up by some huge corporation and hopefully the fourth and fifth generation of Feeneys will carry on.

To describe Tom and Bernie, it can be said that Tom was a physical person, pragmatic, who felt that being a joiner wasn't important, and hating paperwork. Bernie had the mind of a bookkeeper, cerebral rather than physical, obsessed with detail, belonging to many organizations that was important to him. I believe their personalities complemented each other and were vital to the success of the business. Although they both loved golf, they never played together.

About the Feeneys and their boat operations, one could add an assortment of other data, some of it routinely informative and some actually quite interesting. I uncovered lots of trivia along the way. From Bernie's Towing Book and his Construction Book, here are some examples to shed further light on the fascinating Feeneys of Kingston.

TRIVIA

From Bernie's Towing Book:

"The tug JOHN E. MATTON towed 444,439 barrels (18,666,679) gallons of oil to Canada from 4/7/51 to 12/1/51. The tug used 640 gallons of lube oil and 136,179 gallons of fuel and ran the engine for 3871 hours." I have no idea why Captain Curtain would tell him or why Bernie would want to record it. I'm sure that Ralph Matton would have been surprised and more than a little upset to find that one of his employees would give out that kind of information.

"Dwyer barge #105 is the worst towing boat on the canal. It is one mile per hour slower than anything else."

"The tug DENDEL working for McLain left Columbia Street, New York, with four loaded McLain barges on 8/27/40 at 9:30 a.m. Passed McLain's office at noon and passed Kingston Point at 7:30 p.m. on 8/28/40." (For some reason, Bernie thought this was significant to record!)

"There were 275 vessels frozen in the ice in the canal during the winter of 1936-1937 as follows:

184 Wood barges
 38 Wood tugs
 26 Steel tank barges
  8 Steel motorships light
  5 Steel motorships loaded
  5 Steel barges
  5 Steel tugs
  4 Wood steamers

In 1938, he recorded the largest grain elevators in the country. (One is important to this book.) The Cargill Grain Elevator at Albany, New York, with a capacity of 13,500,000 bushels, was the largest in the world.

In the winter of 1933-1934 and for many others, the Feeneys were able to kill two birds with one stone. They had ten barges full of soda ash tied up at the Grasselli Chemical Plant in New Jersey. They had a place for winter storage for the barges in addition to which they were paid demurrage for each day the barges were not unloaded. The time ranged from three to 147 days, making it a tidy sum for them.

I was pleased (as an Industrial Engineer) to discover that the Feeneys used time studies as a basis for estimating the costs of boat construction. They also calculated the cost of electric power for operating their power tools. To be this precise was rare in the 1920's and 1930's for small boat-builders. They added welding data after they started building steel boats.

The following is a tiny portion of the data loaned to me by the Feeney family.

Welding – A man getting $1.00 per hour plus the cost of rod and power comes out to 9 per foot of weld. On ¼" material a man will do 34.5 feet per hour. On one of our steel decks we would need about 300 pounds of rod.

CHAPTER 13

KEHOE OF BROOKLYN AND LATHAM

en·tre·pre·neur: *n.* A person who organizes and manages a business undertaking, assuming the risk for the sake of profit.

The model for Webster's explanation could have been the subject of this chapter – Captain Martin (Marty) J. Kehoe.

The name wasn't always spelled "Kehoe." It began in Ireland as *Kough*. Marty's grandfather was Philip Kough, a fisherman who lived in Newfoundland in the 1880s. On the wall of Martin Kehoe's living room hangs possessions of which he is most proud. These are a certificate and a medal from Queen Victoria to his grandfather, Philip. The citation reads as follows:

> The ship OCTAVIA during the prevalence of a dense fog, got in proximity to a dangerous reef at Burnthead, Newfoundland. Kough (the assistant lighthouse keeper) hailed the ship from the cliff and told them to heave a line ashore, which was done accordingly and caught by him; he then tied the hawser to the cliff, divested himself of the heavier portion of his clothing, fastened a line around his body and went through the surf. He was instrumental in rescuing nine persons.

The Silver Medal was unanimously voted to Kough. The rescue took place on August 4, 1883. Philip had left his heirs a proud heritage.

The next member of the family, chronologically, is Marty's father, James Kehoe. By the turn of the century, the name had been changed to "Kehoe" and James immigrated to the United States. He first came to Boston and then settled in Brooklyn, New York.

As his father before him, he brought his fishing skills along – first working on a fishing vessel and then owning his own. At first, James Kehoe successfully operated out of Island Park, Long Island. His next base of operations was the Mill Basin section of Brooklyn. By the early 1920s, James Kehoe was into the steam lighterage business with two boats he had converted to tugs. In addition, he had a coal business at the same location. A large part of his business was hauling rap-rap [broken stone for foundations or revetments] for East Rockaway Inlet and Jones Beach.

One of the converted lighters became the tug KEHOE BOYS, and that is where Marty and his brother Clayton began as deckhands. Marty's business education began even earlier.

As a young boy, Martin didn't play sandlot ball or any other sports. At the age of ten, he was spending all his free time tagging after his father. His father was, even then, teaching Marty the business. Sometimes he would have Marty running along the beaches and boardwalks of Coney Island and Manhattan beach with a pair of binoculars trying to locate their boats. Later, Marty had a very poignant comment about this time of his life. He said it was like there was a race on between him and his father – that somehow he had to get going before his father left this earth.

The race ended in 1938 when his father died. By then, the Kehoe business had failed. The "Great Depression" was a factor, of course, but Marty also felt that there was a lack of competent, trustworthy employees. That may have been the case, but one problem in this area was probably not in the lap of the Gods, but in the hands of James Kehoe. Marty had another explanation: "My father didn't finish drinking until he was 60 (when he died) and I got finished when I was 45." This was his reasoning.

Marty Kehoe made one other comment which could be construed as an inflated ego – but then again, he may have been speaking simply with extreme confidence in himself. He said if his father had him when he was younger, "The way I seem to have gone – things might have gone differently." He was referring to his father's business. This does seem to me a statement of confidence and the reader may feel the same way. At any rate, at the age of 20, with $500 in his pocket, Marty Kehoe was on his way.

His first and only job working for someone else was for Henjes Marine located on Cropsey Avenue in Brooklyn. It was run by Gerd H. Henjes, whose family had been in the coal business for over 100 years. Call it fate, *karma* or whatever, but it was doubtless the most significant decision Marty Kehoe ever made. Gerd Henjes was to become his mentor. Marty's education was two-fold. By day he was learning how the business operated, and at night he was working as a crewman on Henjes tugs. Before leaving Henjes, he earned his Captain's license. If we can call his education with his father "high school," then the time spent with Henjes was his college.

Some of his philosophy of operating today could well have come from Ralph Matton and Ben Cowles. A couple of things come to mind (1) Cowles said buy old hulls and put them in shape, for an old boat in shape will earn as much money as a spanking expensive new one, (2) Matton said never hire work done or have something made that you can do yourself even if it's not as perfect as "store bought" – operate as lean as possible.

The tug GERD H. HENJES was bought in 1940 while Marty was working for Henjes, and Marty helped with its conversion. An old steam tug was converted to diesel. There was a device that started out as a joke, but proved invaluable in handling lines. It consisted of a Chevrolet automobile engine – in the engine room – which drove a small capstan on deck through a reduction gear made out of a worm drive rear end taken from a truck. The assembly used a Cadillac radiator for cooling, had a self-starter and a Pierce Arrow automatic governor and employed a regular four-speed transmission and clutch. Varying strains on the towing hawser are taken by the automatic governor and the entire unit was built at a fraction of the cost of a regular capstan.

Marty had two other teachers during this period. They were the lawyers for his father and Gerd Henjes. Although Gerd Henjes died in 1943, Marty stayed on until 1944. He did not see eye to eye with Henjes' son and left Henjes in March of that year.

It was now time for Marty to graduate from the college of practical experience and put all his knowledge to work for himself. Of course, to give this chapter a real Horatio Alger flavor, Marty ought to have to make his first move on a shoe string, and this he certainly did. He bought out Dillion Towing and agreed to pay for it out of company earnings. Apparently his reputation with Henjes satisfied the Dillion Company. For the first time, Martin Kehoe now was in business for himself. He had two tugs, the JOSHUA LOVETT and the TRANSFER 3. They were steam tugs, both hired out to the Atlas Tank Cleaning Company to furnish steam for tank cleaning in the New York harbor area.

With the end of World War II came the opportunity to take advantage of the surplus government tugs that became available. To be a successful entrepreneur, several things have to come together. An opportunity can exist, but it has to be recognized. A buyer is needed for the service or product and the main element is, of course, money. To pull all this together requires organizational skills, foresight as to the possibilities, and the nerve to plunge in. To do this is an accomplishment to be proud of, but the real inner glow comes from pulling it off using little, if any, of one's own money. So, for the second time, Marty reached for and snagged the brass ring. There was a small trade off. He would have to give up working as tug captain even part time. I don't feel this was any great sacrifice for Marty. To me, being a crew member was just a means to an end for Marty.

He now had the knowledge of the tug boat and of towing operations from the crew's viewpoint, and the business aspect from his tutelage from Henjes, his father, and their respective lawyers. There is one more key element not yet mentioned – contacts. Contacts cannot be over emphasized. Without contacts, all the rest is an academic exercise. As an example, consider Marty's long association with Mr. Frank Belford, a vice president of the largest towing outfit in the United States – Moran Towing Company.

In 1946, Mr. Belford helped Marty buy two surplus tugs and arrange a contract with Moran to keep them working. Mr. Belford's man picked out one in Madisonville, Louisiana, and the other one in Wilmington, North Carolina. Marty never saw the tugs until they were brought to New York. Each tug cost $80,000 and had to be modified before they could work the New York State Barge Canal. Their wheelhouses had to be cut down or they wouldn't fit under the bridges. A special removable house was built so they could work in the winter in New York harbor. This was not a new idea. Ralph Matton had done the same thing on some of this tugs. Although this type of wheelhouse was a nuisance to install and remove, it was considerably cheaper than any form of power operated up and down house.

The work was done at Jacobson's Shipyard in Oyster Bay at a cost of $25,000 for each tug. The next question is obvious. Where did Marty Kehoe get $210,000 to start this adventure? He formed two corporations and borrowed the money from Marine Midland Bank on a Moran Company signature. He was proud of the fact that he opened each account with only $20. At this time, Marty moved from North Pier Street to the foot of Hamilton Avenue, also in Brooklyn.

The charter with Moran said that they would do their best to keep both tugs working and for that they would take 10% of the gross receipts. This turned out to be a rather short-term contract. After one year, Moran could provide work for only one tug. Marty was given a release and sold the MARGARET KEHOE to Morania Oil. Two years later, in 1949, the CLAYTON P. KEHOE had to be sold because of lack of work. It went to Lake Tankers Corporation for $140,000, the same amount Marty received for the MARGARET. Most people would have been lucky to come out even in circumstances like this. Marty wound up with a small profit.

Even with the loss of two tugs and the Moran contracts, Marty was still in business. In 1947, he had purchased the tug MARY O'RIORDEN and in 1948, the PROSPECT 2, which he renamed the second MARGARET KEHOE. Both of these were used in canal work. This was a decision year for Marty. From this time on, he would specialize in canal work.

A chance meeting at a conference at the Whitehall Inn in upstate New York led to his next business adventure. The man he met was Dana Bray, a CALSO (now Chevron) distributor from Burlington, Vermont. At the time of their meeting, Morania Oil of New York had the towing contract with CALSO to haul fuel oil to the Bray terminal. It didn't take Marty long to convince Dana Bray to dump Morania in favor of himself to haul his product.

Now that a deal had been made, Marty had to buy more equipment. He bought the tug CHOCTOW and the barge HYGRADE 18 from the Bushey organization in Brooklyn. They were renamed the COLLEEN KEHOE and the ANNE BRAY. He also bought the barge SEABOARD 44 from Moran and changed it's name to the DANA BRAY. The partnership came to an end in 1972. At this time, Marty's brother, Clayton, negotiated a deal with Dana to haul his product at a lower rate than Marty was getting. From that time on even though Bray held stock in Marty's companies, he was a competitor. Marty bought out Bray's stock in 1977. There is some poetic justice in that Marty took the contract from Morania and then his brother took it from him. Marty later said that he was probably the only one to have dealings with Dana Bray who came away with his whole hide.

While tracking down all the boats that Marty owned, I ran across a small wood scow named LITTLE BILLIE. I wondered what he would use a wood scow for. The water in front of the Bray terminal at Fort Ann was only four feet deep when they built it. It was unusable until the State dredged it to 12 feet. Now they could bring the barges in but had no place to dock, hook up, and pump out. To build a dock would be very expensive, so Marty thought of using a small scow anchored to shore and mounting the equipment on it. That worked well and is another example of Marty's thinking. Do it quick and keep costs to a minimum. It was during this period with Bray that Marty moved again. He moved his base of operations to Bushey's yard at the foot of Bryant Street in Brooklyn. This move was forced on Marty in 1959 because the land his was on was acquired with the building of the Brooklyn Battery Tunnel.

As I listened to my tapes and recalled the interviews with Marty, I tried to find words to describe him. Tall, thin as the proverbial toothpick, he neither drinks or smokes, and I sometimes wonder if he ever eats. With no hips, it's a guess as to how his pants stay up with no belt to support them. Every few minutes, he has to hitch them up. That may be a nervous gesture in the same fashion as his habit of constantly running his hands through his hair. His Brooklyn accent comes through loud and clear as it does with his daughter, Maureen. He has trouble relaxing. He seems to thrive on crisis.

With a big city background one may develop a suspicious mind about people you don't know well. I speak from experience. I lived in New York City for a time, and I felt the same way until I moved back to a small community. Marty was surprised that I didn't have a business card and letterhead stationery. Fortunately, my first contact with Marty was in the phone, and I was successful in convincing him that I was a legitimate researcher. I still balk at using a business card, But Marty told me that a letterhead paper would help open doors and add credibility to my mission to write a book.

Early on, I noticed Marty played everything close to the chest. I think I know where this comes from. As a loner in a very competitive business, he must keep everyone off balance. His fuzziness of speech sometimes is a function of survival, much like that of a politician. One time he would bare his soul when I least expected it, yet, at another time, he would evade an innocuous question.

Marty is very proud and vocal, as well he might be about his successful survival against some large competitors. Part of his secret is operating with a lean staff. He has operated as many as nine tugs with himself and one person in the office. A more important ingredient is his uncanny ability to underbid his competition by fractions of a cent per barrel. Two examples come to mind. One year Bushey, Marty's chief rival, bid 98.7¢ per barrel to haul jet fuel for the Plattsburg Air Base. Marty bid 98¢. The next year, the bids were even closer. Bushey bid 94.5¢ and Mary bid 94¢. Actually, as Marty himself admitted, the bids realistically should have been in the range of $1.20 per barrel.

I expected to find Marty on the waterfront in Brooklyn. Instead I found him in Latham, New York, a suburb of Albany. It seemed a strange place from which to operate a towing company. Later I learned he had worked out of places much more remote. My first contact was by phone, and I expected the usual short conversation with a busy businessman. I told him briefly why I wanted to visit him and asked if we could set a date. He agreed and kept on talking for twenty minutes. I had hit pay dirt. Here was someone not only knowledgeable, but a great talker.

Seeing his office for the first time was an eye opener. I thought I would find Marty in a shed on the banks of the Hudson River, somewhat like he had at Bushey's yard. His "office" turned out to be a rather plush apartment in an upper middle class complex miles from the Hudson River. Two rooms were set aside for offices, a tastefully furnished living room, a kitchen, and bedrooms. Most of his business is conducted over the phone. He seems never to be more than a few feet from one of them. He has six in this apartment – instead of an assistant, he uses his telephones. Each of his tug crews call collect every day. He calls all over the country the way you and I might call the folks next door.

Of course there are times when business can't be conducted over the phone, and at those times, Marty will travel with great reluctance to New York where his current partner, Carl Ecklof, lives. This reluctance surprised me. The cliche about born and bred New Yorkers popped into my head. The one that states civilization ends at the George Washington Bridge. Anything beyond is really the boondocks. Marty stills owns a house in Valley Stream, Long Island, but hates the area. He dreaded the drive from there to Brooklyn each day. Apparently his wife and some of the children don't like it any better because they live in Florida. I asked him if it was difficult being separated from his family, and he answered with his philosophy of life: "You're no good to anybody unless you make a living – business comes ahead of everything." He said he and his wife have been together the equivalent of four years out of 27. It seems to work for them.

Dana Bray, besides being his partner, Marty said he was a good companion with whom he liked to pal around. Like Marty, Dana was an entrepreneur, always looking for a deal. The two men were linked by two deals that had nothing to do with tugs and barges, but are relevant to a consideration of Marty's character. Marty stumbled into the first deal in Miami Beach in 1958. He was staying at the Treasure Isle Motel on the causeway between Miami and Miami Beach. In talking with Marty, the manager mentioned that he'd like a motel on the beach rather than one on the causeway, and asked if Marty would be interested in buying him out. A call to Dana Bray and a quick trip to Florida and Marty and Dana were in the motel business.

Marty remained in Florida to run the motel, thus beginning a period that would run until 1967, during which he operated his boats by remote control, as it were. He had now built his fleet to eight tugs and three barges. His landlord at this time was the Bushey family. Perhaps that seems odd, since they were competitors, but they were nonetheless friends since childhood. Besides, their fathers had been friends for years, and there were other connections.

At times, Marty pushed oil barges for Bushey, bought one barge and six tugs from him, and after the Liberty Shipyard closed, gave Bushey five million dollars worth of repair work. This lovely relationship, however,

Capt. Marty Kehoe solving problems. Engineer on one phone, tug captain on the other.

Kehoe docks—Mill Basin, Brooklyn, New York, 1930.

Young Marty lends a hand, father James on right. Circa 1920.

Mary and his dad, 1930's.

Example of a "pigeon coop" wheelhouse.

Kehoe fleet, Court Street, Brooklyn, New York.

Paul Gordon, Chief Engineer, surveys damage done by tug ramming a bridge.

came to a halt in 1976 because Frank Bushey didn't like Marty's new partner, Carl Ecklof. Bushey raised the rent up to a point beyond what Marty would pay. It was time to move on again, and that's how he came to be in Latham, New York.

While still based in Brooklyn, Marty called each day from Florida to his one office employee, Catherine Long. Managing the business this way wasn't easy, and at least once a month, he had to come to New York to deal with problems that could not be handled on the phone such as banking, overhauls, and of course, emergencies – which are a large part of this business. This is especially true when you are running old boats. Except for the two war surplus tugs, the average age at the time of purchase was 39 years. In fact, the MARY O'RIORDEN was 84 years old when Marty bought her. Machinery breakdowns and the difficulty of finding spare parts can be a nightmare.

Anyone who knows enough about the day-to-day problems in the towing business has to be impressed that Marty was able to work this way for such a long period. A lot of credit also goes to Catherine Long. One other element that made it possible for Marty to operate long distance was the knowledge and loyalty that resided in the crews that worked for Marty.

A significant change in towing was made in 1965 – double tows came in. Several companies had tried this and gave up after a short period, but Marty ran a double tow for five years, far longer than anyone else. How was he able to do this? His secret was in having a steady captive business. His partnership with Dana Bray provided that. This meant he was able to load and unload both barges at the same location and same time. Others who tried double towing didn't have this luxury.

The object of double towing was to push more oil in less time and increase profits. Marty's unit was the longest unit that would fit in the Federal Lock at Troy. His tug, JAMES J. KEHOE (formerly CHOCTOW), was 77.5 ft; the lead barge ANNE BRAY (formerly HYGRADE 18) was 208.5 ft; and the DANA BRAY was 195 ft. Total length of the tow was 480 ft. The double towing ended in 1970 when Marty's brother, Clayton, underbid him for the Bray contract.

Marty's chief rival, Frank Bushey, chose a different angle – to make a jumbo barge. He took a 17,500 barrel barge and stretched it so that it would hold 25,000 barrels. These units had to be double locked so they had a bow thruster engine which made their total length 297.75 ft. For the sake of comparison, the jumbo barges carried 6,000 barrels less than Marty's two barges, and the difference increased to 10,000 barrels when

Marty scrapped the DANA BRAY for the GERMAINE BRAY. But to compare Kehoe to Bushey really isn't fruitful. For Bushey's needs, the jumbo barges worked where they could not effectively use the double tow. Marty really had a unique situation.

Marty and Dana Bray were together for one last time before their breakup in 1972. In 1967, they bought the Thunderbird Hotel in Las Vegas in partnership with other men. Marty served as the executive purchasing director, but this did not last very long, for nine months later, the Thunderbird Hotel went into bankruptcy, and it was all over. It was now time for Marty to get back to the towing business full time.

It's not hard to understand Marty's decision. While making graphs of Marty's tug ownership, I saw that he had operated the most boats during the periods he was physically away from Brooklyn. Marty admitted that even with one tug you have to be on the scene. Trying to run eight or nine tugs long distance was not good. It took him ten years to get his company back to where it had been. During those years, he was running an average of five tugs and no barges. In 1977, after 33 years in business, Marty finally was debt free.

In 1970, an excellent partnership between Marty and Carl Ecklof of Staten Island began. The two men first met when Carl came over to Bryant Street to watch Marty's equipment being overhauled. Although Carl owned a few tugs, his main business was operating barges. While they didn't meet at the time, their relationship really began in 1968 when Marty chartered a barge (which he later bought) from Carl. Then in 1971, Marty sold the ANNE BRAY to Carl. The partnership is still going today, but on a much reduced scale because of economic conditions.

Marty is fond of saying that it is the upstate farmers who keep his boats running. Larry Pauquette is the personification of what Marty means. This deckhand was born on a farm in North Adams, Massachusetts, on June 4, 1925 and his life was balanced between tugboats and farm.

Like many of his generation, Larry served four years in the Navy during World War II. Boating jobs were scarce when Larry left the Navy, so he worked at various jobs before catching on with a towing company. His workload was awesome. He worked full time as a route man for Freihofers Bakery in Troy, New York. His second job was on a dairy farm and part time he sorted mail at a Railroad Post Office. With little sleep and a lot of running, he was rail thin and more than a little dissipated. While delivering bread to Captain Jack Maloney's house, Dorothy Maloney was quick to notice Larry's condition. As Larry said, "Being the nice woman that she is, Dorothy insisted that I come in for a cup of coffee and a short rest."

Jack Maloney was on his time off and he dropped into the kitchen. Noticing Larry's run-down condition, Jack asked what the problem was. "Nothing except working three jobs is starting to wear me down," Larry told him. "If you can't make it on one job, you might as well give up," Jack suggested, adding that Larry go see Margaret Matton about a job on the Matton tugs. Following the advice, Larry landed a job as a deckhand on the tug MARGARET MATTON. A good word from Jack probably had helped.

Less than seven weeks later, Larry was in a private hospital in Dobbs Ferry, New York, with a severed finger and other injuries, the result of an accident on the tug. He had just gotten out of the operating room when Margaret Matton called in and brusquely told Larry to get to Marine Hospital in New York City as quickly as he could. The reason was simple: the Dobbs Ferry Hospital was costing the Mattons money, whereas Marine Hospital was free.

The only money Larry received was five dollars for cigarettes. Larry asked Margaret if she would advance him enough money for groceries for his wife and children as he had no income. Margaret said she couldn't do that, but she assured Larry he would have a job when he got out of the hospital. Well, there was no job waiting for him and Margaret said he should consider himself laid off. Larry lost no time in getting to New York to hire a lawyer. That action got an instant reaction from Margaret. She couldn't have been sweeter when she called to say that they could work things out if he would stop by the office. Larry's predictable answer was "Too late for that, Mrs. Matton." It was six years before the suit was settled in Larry's favor.

Larry did a little better with his next job. He went with Marty Kehoe and has been with him ever since. I had seen Larry's lawsuit papers while I was researching the Mattons, and I never thought I would meet him, but one morning in 1977 on board the ERIN KEHOE, we did.

Larry presented a fearsome sight as he sat quietly in the corner of the galley that morning. I was not too thrilled with the idea of introducing myself. With his full beard and cold eyes, he didn't look too friendly. I didn't realize how big he was until he rose. His head barely cleared the overhead as he turned to face me. A great smile broke out on his face as he stuck out his hand and said hello. He almost turned my hand to mush as he gripped it a as his rock hard callouses bit into my flesh. With his size and beard, he looked all the world like the reincarnation of an 18th Century French-Canadian voyager. I wouldn't have batted an eye if he had muttered "By Gar!," or some such colloquialism.

Larry is a typical canal boatman. He only works on the boats during the canal season, and he has no desire to go steering. I asked him what he liked best about his job. His reply, "Getting away from home (pause), and getting home again." Larry is completely loyal to Marty Kehoe, yet he would quit in a flash if he could make a living working full time on his farm.

He was a fortunate man with his wife and ten children. He renews my faith in the American work ethic and the solidarity of family life. In these days of changing values, Larry still gives a day's work for a day's pay. He has the respect of his employer, fellow workers and his family.

Another person who keeps Marty Kehoe afloat is Captain Bob Gordon's brother, Paul. Kehoe people kiddingly call Paul "Super Chief," but in fact, he is Marty's Number 1 engineer ashore and afloat. Like his brother, Paul's first job was with Matton, but he didn't get the job in the usual manner. Paul had run away from home and was headed for Buffalo where he hoped to find his Dad who was on the boats. Not having much luck catching a ride on a tug, he stopped in Amsterdam (Lock 11) on the Western Canal and stayed with the lock tender for three days hoping his luck would change. On the third day, the H. A. MELDRUM came along and Paul asked if he could ride to Buffalo with them. Captain Herb Fountain said they were only going to Utica, but they needed a deckhand and asked Paul if he wanted the job. That was September 6, 1941, and Paul, just 15 years old, began his working career on the boats.

He stayed the rest of that season and to July of the next season when a small dream was realized. While working for Matton, he was very impressed with the Bushey tugs with their grained (steel painted to look like grained wood) finish. To Paul, they looked like yachts, and he dreamed of working for Bushey. He went on the CREE as a deckhand. At least he now had a foot in the door. There were no oilers at the time and since Paul had no engine room experience, he had to stay a deckhand a little longer. Oilers were the entry level job for the engine room in the same way that deckhand has always been the entry level to become a mate of captain.

After military service in World War II, he found no openings on the CREE in 1946, so he took a job with the West Virginia Paper Company in Mechanicsville, New York. Lady luck must have been with him. He worked only one night (and hated every second of it) when a call came for him to go on the CREE again as a deckhand. With two years out to run a gin mill and a short stint with Connors Marine, Paul's real career started in 1951 as an oiler on a Bushey tug. His banner year was 1955 when he began as the second assistant engineer on the CREE and finished the year as the chief engineer. The following year he went with Kehoe, seemingly a puzzling move. Having just made chief in a large, well-run company, why quit? But Paul's answer is the same as many others have found. "It is much better to be with a company where you deal directly with the owner" he told me, "than going through foremen and yard superintendents."

As the years went by, Paul gained experience and the trust of Marty Kehoe. He was given more freedom to make decisions and was mostly on his own. After eight years, in addition to being the chief on the ERIN, he became Marty's shore engineer. This meant he was responsible for maintenance of all the Kehoe boats. As the chief troubleshooter and overhaul boss, he is kept busy all winter.

His job is extremely demanding. As an example, he might be on the ERIN in Albany or New York when the JAMES breaks down in Plattsburg. First he tries to solve the problem by phone. If that fails, he has to go to the boat and spend whatever time is necessary to get the tug running. If he figures the time away from the ERIN will be short, an assistant engineer will cover both watches until Paul return. If it looks like the job will take more than a day or two, Paul has an engineer on his time off come on the ERIN until he gets back.

When I asked Paul what area he has the most trouble with, he said problems with the electrical systems. Don't jump to the wrong conclusion, for the electrical problems have nothing to do with the age of the tugs. Paul spent thirty minutes on the subject of incompetent engineers with little knowledge of engines and less on the basic of electricity. Some of Kehoe's engineers are watch standers, but not equipment fixers. Engineers on small tugs are not required to hold licenses, although a few do. As long as there is a Paul Gordon or a Dave Oliver around the lesser lights can slide by. One of Paul's stories provides an example of what he was talking about. There is a way to measure the space between the top of the piston and the cylinder head without half tearing an engine apart. This is done by twisting two pieces of soft solder together and then pushing it into the hole where the fuel injector has been removed. Next you turn the engine over slowly with a steel bar until the top of the piston mashes the solder against the cylinder head. Then the solder is pulled from the hole and its thickness measured with a micrometer. Doing this in front of one of the marginal engineers triggered a truly stupid question. The man asked Paul what he was going to solder down in the hole. The only answer to this is a string of four letter words.

Spare parts was another area of concern for Paul and Marty. It would be difficult to overstate the problem. Sometimes Paul spends $100 on phone calls to locate one part. Someday soon crucial parts won't be available at any cost. At that time, the engines will have to be replaced at a very high price. When Paul is asked where he keeps all the spare parts since Marty has neither boatyard or warehouse, he laughs. The parts are in the basement of Paul's house and every nook and cranny of each of Marty's tugs. There are three areas where the Kehoe engineers do nothing but minor adjustments. All serious problems with the radio, radar, and refrigerators are handled by outside vendors.

I was impressed by the type of jobs Paul tackles that are normally done by a fully manned and equipped shipyard. I recall two good examples: Two of Marty's tugs had their wheelhouses were demolished when they hit bridges. It cost about $15,000 to rebuild the house on the MARTIN and took Paul most of one winter to do it. A shipyard would have charged $50,000-$60,000 for the job. The other example can be seen in the photos of the ERIN. An up and down wheelhouse would have been too expensive so Paul designed and built a fold back upper wheelhouse [sometimes referred to as a doghouse because it was so small] that did the job for a fraction of the cost. These jobs require skill in welding, burning, ship fitting, hydraulics and electricity, all of which Paul has in abundance.

Paul had one complaint that seems to be universal with anyone who does mechanical work. It drives us nuts to have someone hover over us as we work. Paul's cross to bear in this department is his boss, Marty Kehoe. Marty is your better-than-average "Nervous Nelly" who not only looks over Paul's shoulder but offers comments about safety. Neatness of the workplace is not a high priority when tearing an engine apart. It's noble of Marty to worry about his men's safety, but it can sure wear thin if it's overdone.

Pressure and irritation on a job such as Paul's can build up to the breaking point. Paul came to grips with this shortly after the JAMES J. KEHOE sunk at the dock in Cohoes on December 11, 1976. It was not the first time one of Marty's tugs had sunk, and it's a kind of occupational hazard in the towing business. It is not as prevalent today as it was when tugs were wooden-hulled, but it still happens. Anyway, sleeping aboard the JAMES one night after it had been raised, Paul made a decision. He had slept very poorly and kept waking up so he gave up and rose early and decided that from then on, no matter what happened, he would try to laugh at the situation. The pressures of the job have not changed because he still has to keep Marty's aging fleet going, but he does try to put problems in perspective. One fact stands out above all others when we look at Paul Gordon. All his skills are self taught and he is one hell of an engineer.

Although he is no longer with him, Captain Bob Gordon was one of Marty Kehoe's key employees. Without Bob's friendship and tolerance of my lapses in tugboat protocol, a large portion of this book could not have been written. Some of Bob's traits stick in my mind, for example – he never abuses a tug's engines and he always treats a boat as it he owns it. The welfare of his crew always carries the highest priority. He is a cut above many captains in that in addition to the canal, he can run the Great Lakes, Long Island Sound, New York Harbor, and deep water with any type tow. When men talk about top notch boatmen, Bob is in the company of Jimmy Clinton, Jack Maloney, Harry McCormic, and Amos Yell, Jr.

Scattered through out my conversations with Marty were some operational gems and a first that bears repeating. High horsepower, he said, is no advantage in canal work. There is some gain while running in Lake Champlain or the Hudson River, but the added cost of operating larger engines wipes out the advantage of the speed gained in the lake or river. Phrased another way, a 450-horsepower tug can beat a 1200-horsepower tug if less ballast is carried in the barge. You lose time going up loaded but more than make up for it on the trip down empty. When speaking of a first, I have to start with the usual disclaimer that as near as can be determined, the following is one: The only product that Marty sells is time and he is continually thinking of ways to save it. The standard operating procedure after pumping out the barge is to fill one or more tanks with water for ballast. A barge cannot be run empty for two reasons: An empty barge rides too high in the water for the tug crews to see over it, and secondly the barge acts like a huge sail making it almost impossible to control even in a light breeze. There is a penalty for putting water ballast in a product tank. It can't be pumped over the side when it is no longer needed, due to environmental laws, so it costs $600 and time to pump it out at an approved location plus time to clean the tanks.

Marty has had Carl Ecklof (owner of the barges) isolate two compartments in the middle of the barge, one starboard and the other on the port side. These are used for the ballast tanks, and of course, the water in them

248

is not contaminated and can therefore be pumped over the side in any body of water. He has saved precious hours not to mention $600 each trip. This has reduced his product capacity a small amount, but he more than makes up for it in saved time pumping and cleaning tanks at a terminal. Another important point is that training of the crews has to be his way rather than how the Union might prefer it. His people have no stomach for training Union men sent from New York City. Most of Marty's new men are recruited by and related to his existing crews. So far, the Union has avoided any confrontation on the subject. Marty feels he can play it out to the end of the canal with the experienced men he has and if the Union leaves him alone.

How long Marty can survive when one considers that all business on the canal is declining at a rapid pace is difficult to project. His tugs are very old and parts are getting extremely hard to find. There are few young men who are willing to become mates and captains. Finally, the canal structures are deteriorating.

Marty could just quit, but I can't see him doing that. On the other hand, he always hedges his bets, examines all his options, plans ahead, and is not afraid to gamble. There may be a light at the end of the tunnel, so to speak, in his daughter, Maureen. Maureen worked for her dad in 1971 and 1972, left and then returned in 1981. She seems to be the only one of his children who has a sense of pride of ownership and is proud of the Kehoe name. She is intelligent, with an abundance of business savvy.

An example of Marty's planning can be seen in his actions during the winter of 1981-1982. He spent a quarter of a million dollars to put his tugs in tip-top shape. It was a gamble, but if he has to shift from canal work to more difficult water, he'll be ready. Another example of how Marty's mind works is how he deals with his age. He avoids telling his age to his crew because they might become nervous thinking that Marty would retire and there would be no one to take over the business. The crews are breathing a sigh of relief as Maureen is being prepared to take over.

Marty Kehoe is a survivor. Against great odds he has seen many of the big outfits like Bushey, Connors, Matton, Moran, and Russell disappear from the canal. With daughter Maureen next to take the helm, the name Kehoe should shine for many years to come.

CHAPTER 14

SPECIAL VESSELS AND SPECIAL PEOPLE

The SARATOGA and CARSWEGO were unique, the DAY PECKIN-PAUGH is different, the MOHAWK is one of a kind, and the cement barges were a disaster. Then there's the "hoodledashers," which were a hybrid. Categorizing these vessels this way may seem odd, but let me explain.

The SARATOGA and CARSWEGO were unique in that they were designed specifically for canal use and their design had never been tried before. The DAY PECKINPAUGH is different from others of the same design due to its conversion to haul cement. The MOHAWK design is not unique, but it was the only push boat ever used on the canal. The cement barges built for the government during World War I were a disaster not because of a design flaw but because concrete was used for their construction. Their weakness lay in how easily their sides could be punctured.

"Hoodledashers" were something else again – they were the first vessels to combine a power source with space to carry cargo in the same hull. One of the frustrations in writing about canal times is the difficulty one finds in trying to discover the origin and meaning of once-current words and terms. "Hoodledashers" is a case in point. The proper name for these boats was steamers, and they were a natural evolution from animal-drawn barges to today's powerful tugs and motorships.

The bottom line in any business is how to increase profits. Some boatman, whose name is lost to us, sat down one day and figured if he could fit a steam engine and a boiler into one of his barges he could prob-ably tow four, maybe six barges with his invention. It would have been too expensive to build a steam tug, besides with some room left over, he could carry some cargo. In fact, that cargo might just pay for the opera-tion of his new power plant. That is my version of how the "hoodledasher" was born sometime in the 1880s. My friend, Dick Garrity, pointed out an oddity in the fuel that was burned in these steamers. Most coal-burning vessels, big and small, used soft coal in their boilers. The "hoodledashers" used pea-sized hard coal; although it was more costly, it burned cleaner, hotter, and longer which made it a little easier on the firemen.

Although technically it couldn't be called a hoodledasher because of its diesel engine, the FRANK A. LOWERY, for intents and purposes, was one. Originally built as a wood barge in 1917, it was converted to a hood-ledasher design by reducing her length from 150 feet to 110 feet, installing an engine and crew's quarters. She made her last run in the early 1950s and her remains now rest in Roundout Creek at Kingston, New York.

The launching of the one-of-a-kind MOHAWK nearly started a war. Although such boats are quite common on the Mississippi and Western Rivers, the MOHAWK was only the second one in the New York area and the only one to work the canal. The other one is the much larger ROCKLAND COUNTY owned by Red Star, a Bushey Company that has been pushing trap rock barges on the Hudson River for years.

The MOHAWK is small, even for a canal tug. She is 60 feet long and 22 feet wide, with two 500-horsepower GM engines. The MOHAWK was designed to push two 20,000-barrel oil barges. The automated Engines are operated totally from the wheelhouse, which can be raised ten feet for better visibility, and lowered to squeeze under the low canal bridges. This tug, which cost Moran $200,000, was used from 1967 through 1973 in the canal. It could be said that the MOHAWK was a noble experiment but not entirely successful. First, pushboats have only limited application. They are designed to push square-ended barges in fairly smooth waters. Most East Coast towing companies need tugs with more flexibility due to the variety of jobs they are called on to perform.

In day work (8 hours only), the MOHAWK could be run with a captain and one deckhand. For 24-hour work, she was designed to carry only five men, yet she had six bunks! I like to think the sixth bunk was someone's idea of hedging a bet, otherwise why put in an extra bunk? It's a good thing they put in that bunk.

Moran wanted to reduce the crew size – an understandable effort on their part. Automating the engines to eliminate one engineer was innovative, and automating a galley to eliminate the cook was revolutionary. Getting rid of the cook on a tug is like taking a teddy bear away from a child, separating a teenager from his car, or the TV from a football freak. The food was frozen and all the crew had to do was pick and choose and shove their meal in a microwave oven for a couple of minutes. The fact that the food had been prepared by a French chef didn't make up for the absence of the cook.

This did just about start a war! It was an issue the Union would and did scream about. What Moran knew but chose to ignore was that the food itself was not the issue. As has been implicit in this narrative many times, the galley is the focal point of what little social life there is aboard, and

Pushboat MOHAWK built by Bushey for Moran.

"Hoodledasher" (wood steamer), circa 1920.

Captain Joe (Shin) Roberts.

Grain barge, 300 ft. long, 43 ft. 6 in. beam, 15 ft. 3 in. depth. Cargo Carriers, Inc. and Cargill, Inc. built by Bethlehem Steel Corporation, Leetsdale, Penna. Contract CE 7609 & 7610, July 26, 1940.

Cargill grain boat unloading wheat. Albany, New York.

Loading paper in Buffalo. Number 105 became the DAY PECINPAUGH.

The DAY PECINPAUGH in the Genesee River, Rochester, New York.

Steel steamer built for U.S. Government during World War I. Notice sled-like bottom, Buffalo drydock, 1920.

"Hoodledasher" at International Flour Mills, Buffalo, New York, 1930.

Last Canadian paper boat. End of an era, Lock 1, Champlain Canal, 1978.

Pilot house down, nearing bridge at Adams Basin, New York.

Pilot house after clearing the bridge.

Under the bridge.

"Oops" time and too many coats of paint make for an embarrassing situation for Capt. "Shin" Roberts.

the presence of a cook is absolutely essential. You can't ask for a menu change, banter with, or discuss the ball scores with a microwave oven. Needless to say, it wasn't long before the cook was restored in his position of importance on board, and the sixth bunk on the MOHAWK was occupied.

A good example of what I called "unique" boats can be found among those owned by Cargill, Inc., a very large, privately-owned company specializing in the marketing and processing of agriculture products. Cargill is based throughout the midwest, but because Cargill was a major hauler of grain on the Barge Canal from 1938 through 1962, some detail beyond boat design may prove interesting. Although Cargill has been in business in the midwest since the 1800s, they did not start in New York State until 1935. With the exception of the tug PROTECTOR, they operated with leased tugs and barges until 1939. In their peak year for leasing (1938), they had contracts on 14 tugs and dozens of barges. The first two of eight new types of vessels were built in 1939.

Cargill Grain's unique boats were functional, profitable, and ugly. They were the plain Janes of the canal. One look at the photos will confirm this. The names – not very imaginative at that – were made up of the first three letters of Cargill and the endings of cities in New York State. Examples: CARSWEGO (Oswego), CARNECTADY (Schnectady) and CARBANY (Albany).

All eight of these vessels were built to just squeeze into the locks. They were 298 feet long by 42.5 feet wide, and were twin-screwed but underpowered. The horsepower ranged from 300 to 600. Cargill boats carried two extra deckhands because of the numerous hatches that had to be removed and installed on each trip. The word "lines" is often used to describe vessels. To me, Cargill's boats had no lines, but were simply rectangular metal boxes. They were an integrated tow that consisted of three cargo units and an equally boxy power section rigidly tied together with cables and steamboat ratchets. It was no surprise to discover that these unique boats were not designed by a naval architect, but as it turns out, Mr. John H. McMillen II, Board Chairman of Cargill, Inc., was the designer.

Cargill's fleet was loaded at Owsego and carried their cargoes to their 12-million-bushel plant at Albany, New York. They averaged over 500,000 tons per year. There were a couple of problems with these boats. They were not ballasted down so they would have to tie up in winds that others could ignore. The power units could not be operated very effectively when separated from their tow. There was one feature appreciated by the crews.

The power units were large enough to allow an extra room set aside as a lounge. I recall no other vessels having this feature.

"Different" is the word I used to describe the DAY PECKINPAUGH but she could qualify as one of a kind and maybe even unique. She was built in Duluth, Minnesota, in 1921 for Interwaterways Lines, Inc. of New York and named I.L.I. 101. Five of these motorships were built to haul general bulk cargo between New York and Great Lake ports. The name was changed to the RICHARD J. BARNES in 1926 when the DAY was sold to the Erie and St. Lawrence Corp. of New York. She continued in the same service until sold in 1958 to her present owners, Erie Navigation Company of Erie, Pennsylvania. The DAY was converted in Erie to carry cement and unloaded her first cargo on November 8, 1961, at the Mohawk Valley dock in Rome, New York. It took just ten hours to off-load 1500 tons of product and the entire operation, after hose hook up, required one crew member.

A two-and-a-half-yard scraper operates along the entire (254 feet) length of the cargo hold. The scraper pulls the cement forward, up an approach slope to a spot where pumps take over and blow the cement to shore site silos. Loading time is impressive. When conditions are right, the DAY can take on 1500 tons in as little as 90 minutes. Unloading is a much slower process because the weight of the cement packs it down. With the aid of an auger and air pressure, it is reduced to dust which then can be blown up to the silos. At its other unloading location (Rochester), the silos sit on top of a sizeable hill. Unless you have witnessed the unloading at Rochester, you might well doubt that cement could have been blown that far uphill. It isn't necessary to paint a picture of what cement dust does to this vessel. It is a difficult, never-ending task to keep her clean. It is one of the two reasons that it is hard to attract top-notch crews.

The DAY is different in another way. Her crews do not belong to Local 333 of the New York Union, as all others do that work the canal. The DAY's crews belong to a Great Lake union which is structured on the pattern of deep water sailors unions in that they carry three crews on the DAY. Even the job titles are different. Instead of deckhands, they have able and ordinary seamen. There are other differences such as pay and grub. The officers are paid a yearly salary as opposed to the hourly rates used for canal boatmen. One appealing difference for the crews is unlimited money for food. With the exception of Cargill Grain, no canal tug or barge crew ever was that fortunate.

The other reason is that it's difficult to attract good crews in terms of employment. Like the days on the canal prior to the Union in 1936, the men of the DAY sign on in April and are discharged in December, with no scheduled time off. The few that live close to Rome such as Mate Dan Ronspees of Cleveland, New York, can squeeze a few hours at home while they are unloading. There is some consolation for no time off in that the crew work four hours on and eight hours off, while canal tug and barge crews work six on and six off.

The DAY is unique in two other ways. She is the only vessel I ever saw that carries flower boxes. With tender loving care, one member of the crew tends the flowers planted in two metal boxes welded to the stern of the DAY. The piece de resistance would have to be the boat's Christmas Tree. A real tree is decorated, lighted, and placed amidships in early December. As you might imagine, the Christmas Tree draws many favorable comments from those fortunate to view it.

The SARATOGA is another example of a unique vessel. The SARATOGA was one of twenty steamers built for the U.S. Government in 1919 and 1920 at a time the Government had control of New York State canals. These vessels were 150 feet long by 20 feet wide, twin-screwed oil burners rated at 400 horsepower. Since they also carried 350 tons of cargo, it could be said that they were another version of the "hoodledasher." The drydock photo shows the most distinct feature of these steamers, their sled-shaped bow.

Captain Archie Thurston told us that they were very difficult to steer and the overall handling left a lot to be desired. Dick Garrity was once an engineer on these boats, and he had another complaint. The oil tanks were not cross connected he said, so that you could not pump from one tank to the other if the need arose. Garrity cited an example where he had to transfer oil using a bucket, carrying it from one side of the boat to the other. It was a slow and tedious chore. These boats, all named after New York State counties, had a short life on the canal. Shortly after the Government gave control back to the State, they disappeared, some winding up in the Philippine Islands.

Three barges qualify as special. The first and oldest is the T & S No. 5 built of wood in 1923. The solid bottom was replaced with wire screen so that live eels could be carried from Canada to New York City where there was a heavy demand for them. I recall an incident from my childhood involving one of the eel barges, when the bottom fell out of a loaded barge in the Federal Lock at Troy. It was quite a sight to see a lock chamber filled with a million slimy, squirming eels. It tied up the lock for quite a spell until a clam shell crane could be obtained to transfer the eels to another barge.

The ARTEMIS was the largest and most sophisticated barge to travel on the canal. Named ARTEMIS after the Greek goddess of purity, she is basically designed to carry chemicals and can transport many other products including edible oils and liquid sugar. The barge can transport any product requiring the highest purity quality control. She was built in Beaumont, Texas, for Seaboard Shipping Company, a division of Moran Towing and Transportation Company. The ARTEMIS is 300 feet by 43.5 feet with a capacity of 38,000 barrels. "Active" or "positive skegs" of the Schottel design permit her to be maneuvered electronically either while being towed or when a tug is pushing her. Her "rudder" propellers are powered by 12V-71N General Motors diesel engines. Another special barge is the MORANIA 200, commissioned in 1966 for the Morania Oil Corporation. It is used to haul bulk asphalt. Morania pioneered and specializes in the transportation of hot asphalt.

Throughout the life of the canal, progress could be measured by the increase in size of many of the vessels. The goal, obtained slowly, was to increase the size to fill the lock chambers. A good example would be the increase in the size of Mobil Oil's motorships. Starting in the 1920s, there is the SOCONY 77 – 143 feet carrying 175,000 gallons. Next the ALBANY SOCONY – 195 feet carrying 375,000 gallons. The newest is the MOBIL CHAMPLAIN – carrying 1,148,364 gallons, which misses ARTEMIS's 300 feet by just two and one-half inches. She did not start out as a motorship. She was the barge MOBIL 50 until 1966 when she was reconstructed into a jumbo sized self-propelled vessel.

Over the years, tugs have ranged in size from 50 to 85 feet. The great change in tugs was not in size but in horsepower, from as little as 75 to over 1600. There was one exception in size in the 1930s, when a tug the size of an ocean-going tug rode the canal. As far as can be determined, the SYOSSET at 102 feet, 6 inches, was the largest tug to operate regularly in the canal.

There is one last vessel to write about. It's not special because of its size or design, but for the long years of service and from the number of changes of names and ownership. This tug was built in Ferrysburg, Michigan, in 1914. The number of owners that follow may be some kind of record.

BOAT NAME	OWNER
PHILLIP T. DODGE	Unknown
EUGENE F. MORAN	Moran Towing
SCHENECTADY	General Electric
DAUNTLESS NO. 5	Dauntless Towing
CATHLEEN E. MORAN	Moran Towing
MARY D.	Diesel Tug Roslyn
CLAYTON P. KEHOE	Petrol Transport
DORI	Luria Bros.

She met her end on July 24, 1973, after 59 years of service. The DORI was towing in Kill VanKull when she sustained machinery damage due to crew negligence. The vessel was a total loss; it was sold for scrap for $712.

Special People

To further refute my grandmother who was sure I was a lost soul when I went on the MATTON 10 and all the others with their negative opinions about the character of canal boatmen, I offer a portion of the lives of two tug Captains.

If he doesn't hold the record for longevity, Captain Bob McCloskey certainly comes close. Bob was born into a boating family on August 22, 1896 in Kingston, New York. His father worked on the old Delaware and Hudson Canal hauling coal from Honesdale, Pennsylvania, to Roundout Creek in Kingston, New York.

As a young lad, Bob liked to listen to the stories told by the boatmen who hung around the local firehall. At that time, Bob was on his first job of rolling cigars at the American Cigar Company in Kingston. That factory job quickly lost its appeal, and the lure of the tugs took over.

On July 20, 1914, he took a five-dollar-a-week job as a deckhand on the Cornell Steamship Company's tug, GEORGE W. WASHBURN. He rose to Captain in a few years and eventually worked for every company I've written about, except Ben Cowles of Buffalo. Bob married in 1922 and raised one boy and three girls. In 1955, his oldest daughter died leaving four children. He and his wife took three of the children and raised them to adulthood. Their daughter, Mary, raised the fourth.

Bob stepped down as Captain at age 65, but he didn't retire as one might have expected him to do. He chose to go back down the ladder to once again take the bottom job of deckhand. He wanted to be a deckhand again because, as he said, there is no worry on that job. Also, running at night and in the fog was getting to him. Bob was really in love with his work, and the word retirement never entered his mind, he told me. Sick when he worked in the factory, but never on the tugs, he felt his excellent health was due to the outside work on the boats.

Bob finally quit at age 78 after his wife died. He had spent an incredible 59 years on the tugs, and died one year after retirement. He was hardly the kind of mean, irresponsible person my grandmother envisioned boaters were. Even after 52 years of married life, every time he left the tug to call home, he would tell the crew he was going to call his bride!

The other Captain I wish my grandmother could have met was Joseph H. (Shin) Roberts of Waterford, New York. "My father was an extremely kind man" his son, Jerry said. "We were very close and we would talk for hours. He was a great man, and I learned to love him very much. He had a genuine fondness for other people." As Jerry told me this, I thought that it would be grand if my children felt that way about me.

Jerry didn't resent his father's being away so much because he had a lovely mother who filled in for both when Shin was away. It was a great occasion when his father was home and his mother always shared him with the children. To be able to spend more time with his family, they saved all summer so that he would not have to work in the winter.

There was another side of Shin Roberts, one of courage and strength. An episode will demonstrate what I mean. It began with foul weather as the Matton tug H. A. MELDRUM cleared Oswego harbor and headed toward the Galloo Islands in Lake Ontario. They were towing their barge on a short hawser, but as the wind rose, Captain Roberts decided to play it safe and let out more hawser, which is normal procedure in this type of situation. What occurred next might never have happened if the MELDRUM had carried some form of internal communication system other than the human voice.

Son Jerry was on the fantail letting out the hawser, his Uncle Buck was in the wheelhouse, and his father was above Jerry on the deck of the after-house. The balance of the hawser was at Shin's feet, and he was in the process of feeding it to Jerry. The tug was moving too fast and Shin tried signaling to Buck to slow down. Buck read the signal wrong and made a bad situation worse by speeding up. There was a lot of shouting and confusion. Shin lost sight of Jerry and thought he was hurt because Jerry also was yelling to get his Uncle to slow the tug down.

Shin turned to look for Jerry and made the tragic mistake of stepping into the coil as the hawser was running out at an alarming rate. The nightmare of all sailors had come true. With lightning speed, the coil caught Shin's ankle, sent his body spinning to the opposite side of the boat, and tore his leg off. With almost equal speed, Jerry saw his dad go flying, grabbed an axe, cut the hawser, letting the barge drift away. It was a miracle that Shin was not thrown to the lower deck or overboard where the outcome would have been death.

Jerry pulled the separated leg out of the boot and discovered it was attached at the knee with a piece of skin the width of a rubber band. Shin's blood was pouring out all over the deck as Jerry grabbed for his belt to make a tourniquet to stop the flow. His father never lost his cool or consciousness. Shin calmly asked if it was gone. As Jerry said, his dad never panicked in any situation.

They tried without success to raise the Coast Guard on their mobile phone as they headed for Oswego at full throttle. Repeated attempts to reach someone in Oswego payed off and an ambulance was waiting at the dock for them. On the way to Oswego harbor, Shin never went into shock; in fact, he calmly told Jerry to have the cook make some coffee and bring him a couple of packs of cigarettes.

At the hospital in Syracuse, a doctor took one look and said it was a simple matter of snipping the skin and patching up the stump. The doctor left the room for a moment and a male nurse asked Jerry if he could say something. With permission given, he told Jerry not to let this doctor touch his father. The nurse said there was a Doctor Belden who had just reattached a boy's leg that had been cut off in a train accident. Jerry regrets that he can only remember the nurse's first name – Francis.

Doctor Belden was reached and he agreed to try to save Shin's leg. The operation was a success. This kind of operation may be fairly common today, but it was very rare in 1957. Shin was 64 years old when the accident happened and he refused to become a cripple even though he was in pain most of the time and he always walked with a slight limp. He did go back to work for a period of time before he retired.

In examining the lives of canal boatmen, I believe they were a responsible and congenial lot as might be found in any other field. Bob McCloskey and Shin Roberts were typical of this breed and not the exceptions. They demonstrate both the character and the courage of those who worked the canal.

CHAPTER 15

LOW BRIDGES AND HIGH WATER

Graphs speak more eloquently than words on the extent of the decline in commercial traffic on the New York State Canals. By any measure – number of boats operated, tonnage moved, or miles traveled – the end seems very close. The overall condition of canal structures is deplorable. The canals could have been improved, they could have been saved, but there is a good chance that such efforts would not have been successful. There was the natural evolution of other types of transportation systems. In addition, the needs and demands of shippers have changed dramatically since the Barge Canal opened in 1915.

Even fighting and winning a battle to stop the construction of the St. Lawrence Seaway would have only delayed the inevitable. The inroads made by trucks, unit trains, and pipelines are too great to reverse.

Many schemes have been considered over the years to improve the canal. They have ranged from the reasonable to the asinine. The canal has been studied to death. To illustrate, I refer to a portion of a report by the United States Army Corps of Engineers generated at a meeting held in Syracuse, New York, on November 30, 1977.

The purpose of the public meeting was to report on the progress of yet another study of the New York State Barge Canal system. In part, the report said:

> The New York District, Corps of Engineers, completed a survey scope study and report on navigation improvements to the New York State Barge Canal System in 1969. Based on that study, it was concluded that the State of New York *did not desire major modification of the New York State Barge Canal System at that time,* and with respect to rehabilitation, the value of the benefits was below the level required to justify Federal participation under the laws and policies existing at that time.

> Accordingly, the District Engineer, New York District, Corps of Engineers, recommended, in part, that no further Federal action be taken toward making waterway modification improvements to the New York State Barge Canal System.

This study was authorized in 1973, and the entire study is not scheduled for completion until 1989, the projected cost is $6.27 million. Naively, one might think that would be the end of the study. That's wrong, for the Corps of Engineers proposed budget for 1982 included $300,000 to continue the study!

An example of a grandiose plan is one proposed by Nigel Chattley, a British-born engineer and geographer. He calls his plan ICONN-ERIE. He proposes the canal be deepened to 27 feet and using the dirt removed to create an industrial island off New York City. The 27-foot depth was proposed to match the depth of the locks on the upper Great Lakes. This would make it possible for Great Lakes barges to go down the canal and the Hudson River to New York City. Chattley has spoken of this plan at Canal Society meetings, and he has been written about in newspapers and other publications.

There are, however, negative aspects to his plan. First is the cost, estimated in the tens of billions of dollars. An island eight miles square (as large as Brooklyn) would have an enormous environmental impact on New York City and Long Island. An official of the State of New York said that the proposal probably is technically feasible, but the question, he said, is whether it is desirable.

A contributing factor in the downfall of the canal may have been the union that was formed in 1936. All boat owners I talked with made this point to me. Although a serious effort was made to talk with the officers of the union, it was not successful. The reason can be summed up, I think, in one word: paranoia. Paranoia in the persons of past president Captain Joe O'Hare and the current president. (It is difficult to understand why such officials could be so paranoid. After all, no one is looking to do an exposé or looking for skeletons in the closet. It just seems logical for union officials to provide their input to the running of the Barge System.

Although it seemed that Joe O'Hare might be more inclined to talk after his retirement, this proved to be wrong. One short, frequently interrupted interview with Mr. Cornette, the current president, was all that was accomplished, but he avoided responding to most of the questions.

One doesn't really like to end a narrative on a sour note, but events beyond anyone's control dictate it for any modern study of New York's canal system. Perhaps one positive thing can be extracted by the very fact that this book has been written. As a small, but significant, piece of New York State history, it is important that one's grandchildren read and try to understand the contributions made by all the people who were connected with the

End of season for Kehoe tugs, Federal Lock, Troy, New York, 1968. Photo by Gene Baxter.

Time to close the canal. Federal Lock, Troy, New York, 1968. Photo by Gene Baxter.

LAST YEAR SHIPPED

FLOUR	1950
BRICK	1954
IRON ORE	1955
CANNED GOODS	1956
SOY BEANS	1957
RYE/BARLEY/SULPHUR/PULPWOOD	1958
OATS	1959
COAL	1962
LUMBER	1963
SALT/CORN	1964
SCRAP IRON	1965
WHEAT/PIG IRON	1968
FERTILIZERS/FLAXSEED	1969
PAPER	1972
SUGAR	1976

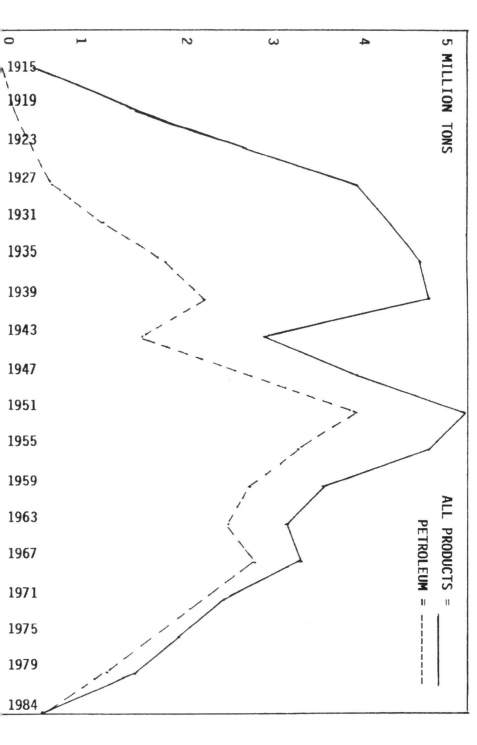

278

canal system. If this happens, then this author's ten-year struggle to research and write the book will have been worth the effort.

Perhaps the end is in sight, simply because the twentieth-century canal was too narrow and too shallow. Besides, too much high water ran under all those low bridges.

EPILOGUE

It's now 1990 and The Diamond Jubilee of the opening of the first section of The New York State Barge Canal. It is also the date that I predicted would be the end of commercial traffic on the canal. I missed the date but the end is in sight. The tonnage for 1989 was less than 500,000 tons.

It is sad to report that Captains' Jack Maloney and Marty Kehoe didn't live to see their stories in print. Saddest of all, after years of support my wife, Helen, isn't here to share this moment with me. I'm not making any more predictions because there are some people that are bucking the odds and may prove me wrong. For the third time, the Oswego River Towing Line is back in business.

Frank & Theresa Coynes' daughter, Evelyn McHugh, and her five children have started the business with a tug I'm familiar with. It was built by the Bushey Company in 1938 and run until sold to Marty Kehoe in 1962. The name was changed to the James Kehoe and painted in Kehoe colors. The McHughs' bought her in 1988 and decided to change the color to grey, name it "Chancellor." I was curious, so I called Evelyn and asked if there was a special reason they used the original name and color. They just liked the name and had no idea that Bushey used the color grey on all their boats starting in World War II. Using bright red trim did improve the looks.

In any event, if hard work, family cooperation and grit is the way to success, then the McHughs' will make it. Good luck to Evelyn, Theresa, Joe, Jim, John and Mike.

ABOUT THE AUTHOR. Charles T. O'Malley was born in Troy, New York, where the Erie Canal and Mohawk River join the Hudson River. His first playground was an abandoned tugboat tied up near his home in Troy, and his first job was firing boilers on an ancient wood steam tug. He served with the U.S. Air Force in the Second World War, and later in the Air Force Reserve and the Navy Reserve, where he held the rate of Chief Petty Officer. He worked for the Eastman Kodak Company, Rochester, New York, for 37 years, retiring in 1985 as an industrial engineering analyst. He is a member of the American Canal Society, Canal Society of New York State, Seaport Museum of New York, Mystic Seaport Museum of Connecticut, and other historical and marine associations. His hobbies include photography, fishing, and, of course, boating. He currently lives in Ellenton, Florida.